CONSERVING OREGON'S ENVIRONMENT

Breakthroughs That Made History

PRAISE FOR *CONSERVING OREGON'S ENVIRONMENT*

"Drawing upon his experience in preserving wilderness areas in Oregon and leading the Sierra Club as Executive Director, Mike McCloskey has the credentials to write this book…in fact, he is the ideal person to write it. Every Oregon conservationist will want to read it."

— GREG JACOB, Professor, Portland State University

"Oregon has long been a leader in the modern environmental movement. McCloskey tells a fascinating story that is little known beyond the state's borders. A great read about an important time in the fight to preserve our environment."

— LARRY WILLIAMS, Co-Founder and first Exec. Director, Oregon Environmental Council

"For the first time, Mike McCloskey has brought together a collection of stories about how Oregonians have long fought to protect the places and values they cherish. If you have always wondered about how so many of Oregon's forests, beaches, and other natural resources were protected, then this is the book for you. It is an invaluable resource for all those who want to understand the background story of those who cared enough to fight for Oregon's great natural legacy."

—RONALD EBER, Historian, Oregon Chapter, Sierra Club

"Mike McCloskey was an active participant in parts of the environmental history that he recounts in this remarkable book. He brings the eye of an astute observer to events that not only were the key to shaping the modern Northwest but that helped form a rising environmental consciousness in our nation. Nobody has pulled most of these stories together in one place before—and nobody has recounted any of them with as much verve and insight as McCloskey."

—JOHN BONINE, Professor of Law, University of Oregon

CONSERVING OREGON'S
ENVIRONMENT
Breakthroughs That Made History

MICHAEL MCCLOSKEY

PORTLAND • OREGON
INKWATERPRESS.COM

Scan QR Code to learn more about this title

Copyright © 2013 by Michael McCloskey
Cover and interior design by Emily Dueker
Smith Rock © James L. McMullen
Mount Hood And Trillium Lake © Bryan Mullennix. Bigstockphoto.com
Map of Places Mentioned in the Text © Martha Gannett. Gannett Design.

Selected items found in this book are from the William L. Finley Papers, Special Collections & Archives Research Center, Oregon State University

Publisher's Cataloging-in-Publication data

McCloskey, Michael.
  Conserving Oregon's environment : breakthroughs that made history / Michael McCloskey.
    p. cm.
    Includes bibliographical references and index.
    LCCN 2013908405
    ISBN 978-1-59299-948-4 (pbk.)
    ISBN 978-1-59299-949-1 (e-book)

  1. Environmentalism--Oregon--History.   I. Title.

GE195.M33 2013   333.72
                        QBI13-600086

All rights reserved. Except where explicitly stated, no part of this book may be reproduced or transmitted in any form or by any means whatsoever, including photocopying, recording or by any information storage and retrieval system, without written permission from the publisher and/or author.

Publisher: Inkwater Press

Paperback
ISBN-13 978-1-59299-937-8 | ISBN-10 1-59299-937-9

Kindle
ISBN-13 978-1-59299-938-5 | ISBN-10 1-59299-938-7

All paper is acid free and meets all ANSI standards for archival quality paper.

3 5 7 9 10 8 6 4

CONTENTS

Acknowledgments ix
Introduction xi

Chapter 1—The Foundational Federal Reserves 1
Origins of the National Forests 1
Crater Lake National Park 16
Re-vesting the O & C Lands in Federal Ownership 20
National Wildlife Refuges 25

Chapter 2—The State and Others Do Their Part 36
Protecting Oregon's Beaches 36
Oregon's State Park System 37
Portland Does Something Important: Forest Park 43
Thwarting Abuses and Mistakes 44

Chapter 3—The Fate of Oregon's Rivers 51
Cleaning Up Oregon's Rivers 51
Keeping Damaging Dams off the Main Stems of Celebrated Rivers:
   The McKenzie, the Umpqua, the South Santiam, and the Lower Rogue 53
The Columbia Gorge 59

Chapter 4—Wilderness Issues in the National Forests 64
Three Sisters Wilderness 64
First Reform in Federal Mining Laws 69
Minam River Valley 69
Waldo Lake Basin: Thwarting Logging 71
Mt. Jefferson Wilderness 74
Wilderness Act Inclusions 76

Chapter 5—Oregon's Environmental Laws: Key Programs 80
Oregon's Beaches Protected at Last 80
Bottle Bill 83
Oregon's Land Use Law 87
Greenways: Willamette 94
Other Greenways in the Making: Frontage Along the Lower Deschutes River .. 96
Combatting Pollution: Air, Water, and Recycling 98
Energy Issues 104
Surface Mining Using Chemical Processing: Oregon Enacts a Path-breaking Law. 110

Chapter 6—Environmental Turning Points 113
Demise of Nuclear Power Plants in Oregon 113
Mt. Hood Freeway 117
Portland Airport Expansion 119
Nestucca Spit Freeway 120
Protecting Eugene's Wetlands 124
Herbicide Spraying 128
Demise of the Elk Creek Dam 132
Portland's Big Pipes 134

Chapter 7—The Advent of New Reserves 138
Oregon Dunes National Recreation Area 138
Hells Canyon National Recreation Area 141
John Day Fossil Beds National Monument 144
Columbia Gorge National Scenic Area 146
Newberry National Volcanic Monument 149
Steens Mountain Complex 152
Cascade-Siskiyou National Monument 154
Zumwalt Prairie Natural Area 156
Marine Reserves 158
Lesser Reserves 159
Reassessing the Oregon Caves National Monument 160

Chapter 8—Protecting Wildlife: Refuges and Programs 162
Willamette Valley Federal Waterfowl Refuges 162
Disputes over Grazing in Malheur Refuge 166
Tualatin River National Wildlife Refuge 167
State Wildlife Management Areas 168
State Efforts to Protect Endangered Species 170
Portland's Heritage Tree Program 171

Chapter 9—Breakthroughs on the National Forests 173
Struggles over Bull Run 173
Growing Conflicts over Logging Roads 175
Adding to the Wilderness System in Oregon 176

Chapter 10—Important Federal Initiatives Affecting Oregon 192
Northwest Power Act 192
Spotted Owl Reserves 194
Courts Manage Salmon 202

Chapter 11—Conclusions 207
Persisting Progress 207
A Place of Importance 208
Innovation 208
Political Parties and Conservation 211
Failures 213
Ease of Acceptance 219

Appendix A—Timeline of Conservation Accomplishments in Oregon 222

Appendix B—Map of Places Mentioned in the Text 226

Appendix C—List of Organizations That Made Conservation History in Oregon .. 229

Index 232
About the Author 248

ACKNOWLEDGMENTS

Various people helped me research the stories recounted in this book. Some, such as Don Waggoner, granted me very helpful interviews. Others, such as Larry Williams, Andy Kerr, Brock Evans, Ron Eber, Sydney Herbert, Griffin Hampson, and Ivan Maluski, told me stories that I incorporated. Some, such as Sallie Gentry, pointed me in the right direction. Others, such as Kay Knack, helped me dig out stories. And for my coverage of Nestucca Spit and the coast, I drew heavily on a draft article written by Catherine Williams.

Others, such as Arnold Cogan and Carolyn Gassaway, reviewed sections. Larry Williams and Ron Eber offered invaluable comments on the entire text. Chris Williams provided indispensable computer assistance.

And I am indebted to Jim McMullen for managing the process of acquiring, taking, and arranging the photographs to illustrate this book.

I thank them all. They made this book possible.

Michael McCloskey

INTRODUCTION

Some years ago I looked into the visitor center for the Redwood National Park. Early in my career with the Sierra Club,[1] I had been their chief lobbyist for the establishment of that park. I was curious about how the Park Service was going to tell the story of that park. To my amazement, I found almost nothing said about how that national park came to be. It was almost as if it were the offspring of a "virgin birth." It had no father.

I knew that this park had come out of a wrenching struggle that drew national coverage. The affected lumber companies fought every inch of the way against the environmentalists, yet now it would seem that this was simply the product of an enlightened government. I then wondered whether the park's establishment was too "hot" a subject for the Park Service to touch.

But as I then looked around the country in the visitor centers for other national parks, I learned that this silence was the rule. Those who had worked their hearts out to bring about these achievements were rarely even acknowledged, let alone thanked. Fortunately, Crater Lake National Park is something of an exception, with the pivotal role of William Gladstone Steel showcased.

For a while I had hoped that those in academia who specialize in environmental history might redress the balance. I had read some excellent books by some of them that told the stories of how such laws as the Wilderness Act had come about. But as I explored further, I learned to my dismay that few of them were focused anymore on public programs. Many of them had shifted their focus from "political history" to what they termed "cultural history" instead. Now they were delving into issues of race, gender, and class. While once these issues had been neglected, they now were their main focus, and public programs were being neglected. Some of them were going even

---

1 I worked for the national Sierra Club for forty years and became their CEO.

Michael McCloskey | xi

further and viewing those who made their livelihoods exploiting natural resources as the victims. Thus, these academics no longer were inclined to view environmentalists as the "white hats."

With Oregon now being widely acknowledged as the leading environmental state, I knew that I had missed a lot. When I retired and returned to the state, I thought that I could catch up on things by merely reading a book on the topic. Since Crater Lake had been something of an exception, I thought Oregon might also be one in this regard. But I could not even find many instructive articles in historical journals. I had to dig the picture out piece by piece. At first, I just went as far as 1970. The Oregon chapter of the Sierra Club showed interest in what I had written, published it, and it was well received. So I went further, to 2012, and this book is the result.

Because there are a large number of people in Oregon interested in conservation, I thought that they would enjoy finding all these threads pulled together in one place. They would not have to search in obscure places to dig out each story. They can either read this book straight through, or use it as a reference book.

This book, thus, aims at telling the story of how the most significant accomplishments in conservation in Oregon originated. I regard these accomplishments as significant for a number of reasons: The areas protected are sizeable, or the actions taken laid the foundations for later progress. In a way, they broke new ground or set important precedents. They have stood the test of time as being valued, or I believe they are likely to.

Obviously, these are accomplishments that have a more or less permanent character. For instance, in many cases the areas are protected by statutory enactments; obviously, the legislative body could choose to repeal these laws, but it rarely does. I have usually not included actions taken just by administrative action. These are too easily reversed as ideologically oriented administrations come and go.

For this reason, I have chosen not to include many matters involving wildlife. All too often, they are regulated by states under shifting regimes. One day's step forward is reversed the next day. Shifting variables, as well as user pressures, push the regulations back and forth. Few of these decisions involve matters of a permanent character. They are also of greatest interest to somewhat different constituencies—in one case hunters/anglers and in another case, those devoted to animal protection.

Thus, I do not try to tell many stories involving salmon, wolves, and cougars. They are told in other books. There are many, for instance, on the plight of salmon and their fate. However, I do tell the story of setting aside federal wildlife refuges. These are set aside under a framework of federal law and are fairly permanent in character. Yes, a limited portion of them are open to hunting, and hunters play a major role in financing them. But they do not exist primarily for the benefit of hunters; they exist to provide habitat for wildlife—mainly migratory waterfowl—where there are treaty obligations.

Traditionally conservation dealt with issues of protecting nature and habitat. When the broader environmental movement began to develop in the early 1970s, interest broadened to also cover energy issues and concerns over curbing pollution. In this work, I cover them all.

While I try to cover a wide variety of issues and those of interest to people all over Oregon, there are some things I have not tried to do. I have not tried to tell the story behind every unit in large systems with many units; there are too many in most cases to make that workable. Usually I tell the story behind the overall system, and tell a few stories behind some of the most representative units. I have also not tried to write a complete history of conservation in Oregon,[2] nor have I tried to relate each story to larger themes at work in society. And I have not tried to relate the stories of failed efforts; e.g., the campaign for an Oregon Cascades National Park. Again, I feel those might best be dealt with in a different work, probably by academic historians.

Mine is a story of efforts that succeeded and that have made a lasting impact on Oregon. They are our heritage. Some are well known, but many are little known. In almost every case, they succeeded because someone or some organization cared deeply about the places or issues involved. Time

---

2 Others have moved in this direction; e.g., see William G. Robbins, *Landscapes of Conflict: the Oregon Story, 1940–2000* (Univ. of Washington Press, Seattle, 2004). A complete treatment would not only tell of the accomplishments of environmentalists but their hopes, efforts, and failures as well.

and time again, these leaders have refused to accept the status quo and made things change. I want to pay tribute to them and make sure they are remembered. We owe them all a lasting debt of gratitude.

I hope you will savor the excitement that arose out of their efforts.

Michael McCloskey

CONSERVING OREGON'S ENVIRONMENT

CHAPTER 1
THE FOUNDATIONAL FEDERAL RESERVES

Origins of the National Forests

Today, almost half of Oregon's forest land is in national forests. Historically, they contained rich timber stands that have been avidly sought by lumbermen, as well as summer pastures sought by those grazing sheep and cattle (especially in eastern Oregon). Over time they have come to be valued for other purposes: for water and later recreation and as habitat. Today they are organized in eleven units, though they have been split and consolidated at various times, with varying names.

There are too many of them to try to tell the stories of the origins of each in detail, but a number originated in colorful contests between idealists, schemers hatching frauds, and pragmatists trying to work out practical ways of managing these parts of the public domain.

Cascade Range Forest Reserve

Most of the lands that are now in national forests along Oregon's Cascades range were set aside as the Cascade Range Forest Reserve by President Grover Cleveland in 1893. At this time, section 24 of the Act of March 3, 1891, authorized executive action of this sort without Congressional action. It grew out of efforts in Congress to reduce the scope of laws then providing for the disposition of the public domain.

Old-Growth Forest in the Willamette National Forest

Running from the Columbia River to the California line, this reserve—at 4.8 million acres—was the largest of its time. It included what we now know as the Mt. Hood, Willamette, Deschutes, Umpqua, and Rogue River National Forests.

It was a year after this Act was enacted before most Oregon residents even heard that the President could take such action. An agent (R. G. Slavery) in Portland of the Interior Department's General Land Office (GLO) revealed that the department had been studying the Mt. Hood area (having already received a request from Portland's Water Commission to withdraw [set aside] the Bull Run watershed) and that they were considering a larger withdrawal. He said that expressions of public interest in federal action would be welcome, inasmuch as he planned to be in Washington, D.C., in

**Judge John B. Waldo in Winter in the Field
April 1916**

the near future and wanted to tell officials what people here wanted. What would they think of a withdrawal, which would close the land to most types of settlement and signal that the federal government intended to hold onto that land?

Since 1885, Judge John B. Waldo[3] of Salem had been pushing to have a large withdrawal along the Cascades, launching a petition drive at that time. When he served in the Oregon state legislature, he introduced a memorial to Congress in 1889 seeking a federal reserve that would include twelve miles on both sides of the crest for the length of the state. While it passed the state House, it did not overcome opposition in the state Senate from sheep men

---

3   He was then the Chief Justice of Oregon's Supreme Court.

who feared they would be shut out of the high meadows. Nonetheless, his effort provided impetus to the idea. He had explored much of the area of the central and southern Cascades on long summer trips, with various natural features, such as Waldo Lake, being named after him. Some even celebrate him as Oregon's John Muir. He reveled in the mountains and wilderness as places of "untrammeled nature and free air."

But there was also another mountain enthusiast in Oregon at that time: William G. Steel of Portland, who had started the Oregon Alpine Club and later the Mazamas. They were friends who worked closely together. When Steel heard of the GLO's desire to sense public opinion, he remembered Waldo's proposal and petition, but there were others who also remembered it. Interests fronting for the railroads and timber speculators saw opportunities to pursue fraudulent schemes when withdrawals were made, and the larger the better. Representatives of these interests approached Steel, suggesting going beyond the Mt. Hood area and petitioning to have the entire range withdrawn.

In quick order (by April of 1892), Steel managed to launch an impressive effort, with a petition in support of this large withdrawal signed by the elite of Oregon and many mayors, and backed by Portland's Chamber of Commerce. The GLO official was then convinced that opinion here was overwhelmingly supportive.

However, by January of 1893 things had become less clear. Steel and his petition signers began to have second thoughts, as they became aware that people with less than clean hands had induced them to broaden their focus. They decided to limit their focus to the Mt. Hood area, and got their followers to sign a second petition along these lines. But in the meanwhile, the Oregon Senate had endorsed the large withdrawal.

In any event, on September 28, 1893, the GLO took action to have the large area withdrawn as set forth in Steel and Waldo's original petitions, creating a forest reserve. In the material they prepared for President Cleveland, no mention was made of their rationale. However, some scholars feel that there is evidence that the GLO was taking steps to limit the opportunities for fraud in the reserve. Knowing what could happen, they may have decided that the benefits outweighed the risks. But soon the GLO found that it had too few qualified staff to survey and fix the boundaries of this reserve and to protect it from trespass.

At that time, the law specified that withdrawn areas were closed to claims and settlers. In 1894 the GLO concluded that the law obliged them to close withdrawn areas to grazing sheep, prompting an outcry in protest from sheep men who used summer pastures there. John Minto, a prominent sheep man, wrote a series of articles opposing the reserve. Oregon's legislature then passed a memorial to Congress asking it to re-open the reserve to grazing (as well as to allow some settlement). Then members of Oregon's congressional delegation tried to reduce the size of the reserve, or even eliminate it; Congress responded by cutting off funding to enforce the grazing ban.

During this time, Waldo and Steel did everything they could to defend the reserve, including unleashing another barrage of petitions and telegrams. At the behest of Waldo, in the spring of 1896 Steel went to Washington, D.C., to lobby Congress. Waldo also sent an appeal directly to President Cleveland. In addition to the Mazamas, Muir and the Sierra Club were enlisted to oppose the efforts of the sheep men, who sought to drastically reduce the size of the reserve.

Eventually, they prevented these bills to shrink the reserve from passing. As late as 1899, Waldo was still having to stay vigilant to block efforts to weaken the reserve.

Altogether, he immersed fifteen years in the project to establish and defend the Cascade Range Forest Reserve.

This imbroglio spotlighted the question of how these reserves were to be managed. Throughout the 1890s, Congress had been trying to enact management legislation, but had not been able to complete the process. The leading bill, which was known as the McRae bill, had twice passed the House, but the Senate would not clear it. The Secretary of the Interior at that time decided not to make any additional withdrawals until Congress decided how they were to be managed, even though petitions from recreational groups were pending with him.

For a moment, Congress thought it might get its answer from the National Forestry Commission, which it funded in 1896 to study the question. Being recommended by the American Forestry Association and operating under the auspices of the National Academy of Science, it had been promoted especially by Professor Charles S. Sargent of Harvard, Gifford Pinchot, and magazine publisher Robert Underwood Johnson (who was close to John Muir). Composed of five specialists and joined on most of its tours by Muir, it conducted lightning inspections of western forests and issued influential recommendations, largely written by Sargent. While it could not reach agreement on all questions, it did recommend setting up a new agency with needed expertise to administer the reserves. The GLO had been largely staffed by political appointees without credentials.

Muir wrote much of its sections on the problems caused by grazing sheep, with which Oregon's John Minto took issue. Instead of heavy regulation, Minto favored a system of leasing pastures and allowing them to be homesteaded.

The commission also recommended setting up thirteen new reserves, totaling over 21 million acres. Outgoing President Cleveland did this in February of 1897—just a few days before leaving office. None were in Oregon. Also he vetoed a bill to eliminate these reserves.

The way for more withdrawals in Oregon was paved finally by the passage of a new organic act for the forest reserves on June 4, 1897. Cleveland's last-minute withdrawals triggered such an uproar in many western regions that a tradeoff was finally possible in Congress. While Cleveland's withdrawals were suspended for nine months (to allow for more homesteading in them), on the other hand the substance of the McRae bill passed as a rider to an appropriation bill. It emphasized that lands more valuable for agriculture should be excluded from forest reserves and that such reserves should be managed to protect and improve their forests, provide a continuous supply of timber, and provide favorable conditions for water flows. It should be noted that W. G. Steel of Oregon had put himself on record in support of the McRae bill.

Reserves in Eastern Oregon Mountains

Blue Mountain Reserve

In contrast to western Oregon where the schemes of timber speculators were the main problem, in eastern Oregon the main problem was conflict between those who maintained grazing herds. In many places, range wars had broken out between the owners of huge herds of migratory sheep and the resident cattle ranchers. For instance, a subsidiary of the immense cattle combine of Miller and Lux had filed on most of the water holes, giving them effective control of most of the pastures. There was even conflict between resident sheep herders and migratory sheep herders. While it was a lesser problem, there also were those who were speculating in timberlands and using dummy filers to claim choice pine forests.

In 1903, a petition was filed for a large, temporary withdrawal in the Blue Mountains. Many of the signers were spurious settlers, having been found hanging around bars. It was suspected the schemers were promoting it to file claims in advance of the withdrawal that could thereafter be exchanged for more valuable lands elsewhere. And there were many more fraudulent schemes.

Opposition arose in a number of communities: Baker City, Canyon City, and in Grant County. They feared fraudulent intent and interference with bona fide homesteading and mining. But those in Prairie City supported it, as did those in at least one county Woolgrowers Association, seeing it as a way to end the range wars. And the local Oregon congressman, John Williamson, also supported it.

Statewide newspapers took opposite positions, with the *Oregon Journal* opposing it and the *Oregonian* supporting it, as did the Portland Chamber of Commerce. The *Journal* suspected too much fraud was afoot.

The Interior Secretary came to be suspicious of such petitions, seeing a pattern of fraudulent intent behind some of them. Having been warned to be careful of being duped, his land agents then proceeded cautiously. Even President Theodore Roosevelt began to show concern, fearing that laxity could discredit the whole process of making timely and needed withdrawals.

On his 1903 trip through Portland, Roosevelt had received warnings about these dangers from Oregon governor George Chamberlain and the *Oregonian*'s editor, Harvey Scott.

State land agent Binger Hermann took these warnings to heart and turned down this petition to make a large Blue Mountain withdrawal. He was particularly concerned about filings that would give too many ranches control of water holes.

But Hermann's action did not save his job. He was soon dismissed for being ineffective and engaging in nepotism. In his place, Roosevelt appointed a former governor of Wyoming, W. A. Richards. Roosevelt directed him to confer with Governor Chamberlain and Harvey Scott about avoiding these pitfalls. And the boundary work was transferred from the Geological Survey to the Bureau of Forestry, along with its chief field inspector, H. D. Languille.

A few months later (on September 7, 1903), the *Oregonian* had a change of heart, running a major exposé of the extent of fraud involved with the proposed land withdrawals. Finding that one-fourth of the state was involved in these temporary withdrawals, it reflected the view that the process was being manipulated for selfish aims. It found the following problems in what it was under way:

- the withdrawals were indiscriminate, embracing many unsuitable areas;
- they were based on hasty surveys;
- much of it was designed to facilitate fraudulent schemes;
- much deplored "in lieu" selections were afoot; these allowed homesteaders who lived within withdrawals to select replacement lands on public domain elsewhere;
- various owners of wagon road claims were trying to benefit unfairly;
- cattle interests were trying to gain control of water holes as part of their base properties so they could gain effective control of vast areas of public land; and
- it alleged that the local GLO was withholding vital information.

The paper did concede, though, that the withdrawals would block the schemes of speculators in timber lands.

Various public officials were accused of being in league with the perpetrators of fraud: for instance, Congressman Williamson and forest supervisor S. B. Ormsby. Binger Hermann was accused of negligence and nepotism. Even H. D. Languille was accused of being in the pay of speculators, though within a year, his name was cleared. He felt he was accused quite unjustly, and historian Lawrence Rakestraw agrees.

Soon the state GLO opened its files to reporters, as the *Oregonian* wanted. In this fashion, they showed they had nothing to hide, but they also may have wanted to shape public opinion since they were about to go to court against the "land sharks."

Acknowledging that some fraudulent schemes had been launched in connection with the withdrawals, state GLO director Richards argued nonetheless that they were the only way to put these tracts out of reach of the timber speculators. The *Oregonian* then accepted this explanation, feeling it had realized the goals of its exposés. Thereafter, its focus was on the trials of the "land sharks" and on its efforts to reform the GLO operations in the state. Some of the evidence of fraud in the Blue Mountains was used as evidence in these trials (see Chapter 2).

Regarding GLO operations, the *Oregonian* argued that in too many cases the people who staffed its offices were political appointees who were in league with the "land sharks." They were routinely appointed on the advice of Oregon's U.S. senators. Called "registers" and "receivers," too often they had passed on critical, official information to facilitate fraudulent schemes.

After one such official had been fired, Senator Charles Fulton tried to persuade President Roosevelt to appoint another of his candidates. Feeling that he wanted to bring an end to this pattern, Roosevelt refused, saying he did not want to give the timber speculators "a free swing." Instead, he insisted he wanted a stronger and better candidate—in fact, "a first-class man." When such a candidate did not emerge from Fulton, Roosevelt turned instead to Governor Chamberlain and to his state land agent, Oswald West, for the kind of honest advice he was seeking.

Once the *Oregonian's* exposé had been made, the Bureau of Forestry turned its attention to the question of what should be done in the Blue Mountains. It worked feverishly to examine the situation on the ground, finding some areas to be unsuitable for a reserve and other areas that ought to be added to its boundaries. Its leaders felt that they were in a race against "land thieves."

Their point man in this work was H. D. Languille, who did most of their work on the ground. Languille began as a climbing guide on Mt. Hood, having guided Pinchot around the area when he visited it in 1896. His father built Cloud Cap Inn there (which was bankrolled by C. E. S. Wood and banker W. M. Ladd). After being trained at Yale, he joined the Geological Survey, doing most of its boundary work in the Pacific Northwest.

Not only did he show an in-depth knowledge of conditions on the ground, he wrote clear, well-considered reports and was outspoken. Historian Lawrence Rakestraw said Languille had "an ardent desire to save the woods of Oregon from eastern lumber syndicates bent on exploiting them."

He endured heroic conditions in the field, though at the outset he had been hurried into approving proposed withdrawals that he had not actually inspected. And then he could be ingenuous in accepting petitions as bona fide that were actually schemes of fraudsters. But Rakestraw thinks his critics were too harsh on him.

In doing this work, Languille felt he was in a race against dishonest claimants, as well as large eastern interests. He saw himself as helping the small, working man and local farmers, who depended on summer irrigation water from these forests. He also felt his own honesty and competence were at stake.

Beginning in 1903, he traveled tirelessly in all seasons to make his inspections and to get to know the areas involved. Travel in winter by horse was particularly taxing. He not only acquired technical information, but he also met with affected local people, explaining the purposes of the proposed withdrawals and reserves. He even tried to win over local miners with the argument that reserves would protect the forests that provided the mine timbers that they needed, pointing out that the reserves would not interfere with their mines and claims. While he noted the beaten-down condition of the area's ranges, which had been lush only a decade before, he chose not to get into arguments with grazing interests whose support he needed.

After hearing him, in many instances hostility died down—as people found out that they had been misled. Languille convinced many of them that the purpose of the reserves was not to facilitate fraud, but to prevent it. Support grew as he moved about with his explanations; for instance, in Canyon City ranchers dropped their protests.

With the new management act, he was now able to argue that it was no longer the case that reserves were off-limits to productive work but instead were designed to facilitate such work by conserving their resources. Generally, real settlers supported the reserves, while the "land sharks" opposed them. Often, even ranchers supported them.

However, Languille was not sympathetic to some efforts made by timber companies to eliminate from the withdrawals forested land that they wanted to log, attempts they were making through Senator John Mitchell. Languille rejected efforts to file claims for these lands because the companies were using dummy claimants and, he felt, really intended to denude them.

The amount of land tied up in temporary withdrawals in eastern Oregon at the time (1903–1906) was immense: five million acres were in legal limbo while the frauds involved in trials were being sorted out. Proclamation of the Blue Mountain Reserve was delayed a year (1905) while the trails proceeded. Finally, they were made on March 18, 1906.[4]

Wallowa Reserve

Seventy percent of this high country in the Powder River basin was forested, though it had been badly scarred by burning to favor pastures. Little agriculture was practiced there; grazing was the main activity, though there was some lumbering. And there was an important mine at Cornucopia.

To forestall timber speculation, Languille made a temporary withdrawal of this area on May 21, 1903. Most of the local public was supportive, with

---

[4] Most of these forests are now in the Wallowa-Whitman and Malheur National Forests.

those opposed mainly in Baker City and involved with mining. A permanent withdrawal of 747,000 acres was made on May 6, 1905. It was combined with a withdrawal of grazing lands near Wallowa City, and called the Imnaha Reserve, and later combined with other lands in the Wallowa-Whitman National Forest.

Wenaha Reserve

This forested, high plateau between the Grande Ronde and Snake Rivers straddled the state line. In Oregon, it included land that lay in Umatilla, Union, and Wallowa Counties. It had not only been badly burned but was over-grazed by 200,000 sheep in 1904, as well as cows and horses.

The GLO had received a flow of petitions asking for a reserve there, including from Lewiston, Idaho, which drew its water supply from the area and sought better protection. Inspector W. H. B. Kent temporarily withdrew it in October of 1902, with some opposition.

But public opinion quickly shifted in favor of a permanent withdrawal and reserve, which was made in 1903. Local farmers saw it as protecting the source of their irrigation water; only migratory sheep herders opposed it. It is now part of the Umatilla National Forest.

Heppner Reserve

This spur of the Blue Mountains is north of the North Fork of the John Day River. Today, it lies north of the towns of Spray and Monument. Shortly after 1900, it supported fine forests, springs in the woods, and valuable pastures, which, however, had been badly over-grazed by thousands of sheep.

A range war had broken out—as herds of sheep from California moved through the area, with hundreds of them being shot. Local opinion felt the cattlemen were justified in what they were doing.

Languille was able to visit the area in 1903 and found that local opinion supported a reserve there to bring order to the range. He had made a temporary withdrawal in May of 1902 to regulate grazing and protect the pasture and the forests. The final withdrawal was for 261,000 acres in

Morrow, Gilliam, and Umatilla Counties. It was supported by cattlemen and sheep men alike. Today this area is part of the Umatilla National Forest.

Maury Mountain Reserve
This range due east of Bend is now part of the Ochoco National Forest and provides protection for the sources of the Crooked River. On lands similar to those found in the Heppner Reserve, an even more serious range war had broken out in the early 1900s. Thousands of sheep were being shot and violence perpetrated, with homes even being burned. Cattlemen were illegally fencing the public domain, and timber theft was widespread.

Languille withdrew the area to bring order to the range and stop speculation in timber—before it was too late. In reaction, public opinion was divided, with sheep men supportive and cattlemen less so—as they then thought they had the upper hand.

Warner Reserve
Range warfare had also broken out in this area in the mountains on either side of Goose Lake. In the early 1900s, this high plateau in Lake and Klamath Counties was used mainly for grazing, although it was forested and burdened by the activities of timber speculators.

Nomadic sheep men drove over 100,000 sheep across this area as part of the California driveway. This grazing was on top of the impact of grazing by over 50,000 resident sheep, cattle, and horses. Large numbers of migratory sheep were being killed in the conflict.

The inspector who earlier had been instrumental in setting up the Modoc forest reserve on the California side of the state line had also suggested that this area be withdrawn for a reserve, but this was not done at that time—despite the fact that the residents of Lakeview wanted a reserve to protect their water supply.

Once the range war broke out, many local residents petitioned for a reserve—having been impressed by the way that the Cascade Reserve had regulated grazing and brought order to the range. There was some dissent, but it was mainly by timber speculators and floaters.

Upon inspection to provide necessary information, a temporary withdrawal was made in July of 1903. The following year, additions were made on its western edge where timber speculators flooded in response to a rumor that a railroad was coming through, which would make it possible to exploit the pine stands. Under the Timber and Stone Act, thousands of spurious claims had been made by dummy entrymen. About ten percent of the area had already been lost in this fashion. Booth-Kelly and Weyerhaeuser stood to benefit most.

With some adjustments to eliminate unsuitable areas, a permanent withdrawal was made in 1906. It drew substantial local support, with tramp sheep men and timber speculators providing most of the opposition. Today, this area is part of the Fremont National Forest.

An addition to this forest was made in 1958 when the Eisenhower administration terminated the Klamath Indian Reservation. To rescue some land from this debacle, Senator Richard Neuberger of Oregon took the lead in having about half of the lands, which otherwise would have been privatized and flooded the timber market, put in the Winema National Forest (formally established in 1961 and now known as the Fremont-Winema National Forest).

Other Reserves in Western Oregon

Siskiyou Reserve

Interest in withdrawing this area goes back to 1898, when the GLO first sent inspectors to check it out. But two of them sent by Binger Hermann brought back conflicting advice. Moreover, the *Oregonian*'s Harvey Scott long suspected fraudulent intent in connection with withdrawals there.

The area was not withdrawn until H. D. Languille inspected it in 1903 and recommended that it be done. He found lots of local support, though timber interests were not aboard.

Siuslaw Reserve

Some of the original Cascade Reserve, in what later became the Umpqua National Forest, was in the Coast Range. And on March 2, 1907, the Tillamook Reserve was established as one of President Theodore Roosevelt's last expansive uses of his presidential authority. Named after the town of Tillamook, it embraced 175,518 acres.

The following year, this reserve was combined with the coastal portions of what we now know as the Umpqua National Forest to constitute the Siuslaw National Forest. On July 1, 1908, a new national forest was set up of 821,794 acres. In time, it became one of the state's leading timber producers, but now it leads the way in environmental reforms of its timber operations.

Not long before this became a federal forest reserve, it had all been part of a sizeable coastal Indian Reservation that was done away with for various reasons—chiefly that it was coveted by white settlers.

Summary of Situation

In this period of 1903–1906, over four million acres had been withdrawn in Oregon. Most of the withdrawals were made in 1904, with public opinion favorable in most cases.

While the President's authority to make these withdrawals was repealed in 1907, Roosevelt had anticipated this and set aside 16 million acres in new forest reserves beforehand in twenty-one new units across the West. Only a few were in Oregon, and the public reaction here was generally favorable. It should be noted, though, that Senator Charles Fulton of Oregon sponsored the legislation that ended the President's authority to establish these reserves.

Roosevelt's further forest withdrawals in 1907 set off another contest of wills between the conservation forces and those who opposed them. Oregon stood out as a state with substantial support for more forest reserves. Oregon's governors at that time, George Chamberlain (1903–1909) and Oswald West (1911–1915), championed them. They were both progressive Democrats, with most Republicans opposing the withdrawals. Oregon's eight-year-long Conservation Commission, set up to find facts on the question, tended to support the withdrawals. With most of its members appointed by Democratic governors, its members, however, reflected partisan loyalties and divisions. In the rhetoric of the time, the issue turned on claims that the withdrawals either prevented monopoly and waste, or that they deprived the state of the tax revenues that would come through private ownership.

However, there were some unusual alliances. Some of the large timber companies supported them, trusting that the Forest Service would do a good job growing timber for them to cut when the time came. They hoped that Forest Service management would bring stability and scientific management to these forests. While sheep men and miners largely opposed them, Oregon sheep men were more divided on the question than elsewhere. Over time, the Forest Service worked hard to placate these commercial interests and retain the support here that it had.

By 1915 the issue had died down.

Crater Lake National Park

One part of the Cascade Forest Reserve did not stay long in the national forest system. A few years after it was established, this part, in the southern Cascades, became Oregon's only national park: Crater Lake National Park. It is generally regarded as one of the "crown jewels" of our nation's system of national parks.

One indefatigable person was primarily responsible: William G. Steel, who had been involved in the establishment of the Cascade Reserve. He spent seventeen years getting the park established.

Steel made his living in unpretentious ways (e.g., working for the post office), but his life's work was really Crater Lake and mountaineering. Living for the most part in Portland, he was the founder of the Mazamas (1894)—a mountaineering group—and the short-lived Oregon Alpine Club (1887). He was in love "with open, wild and beautiful places."

He wrote a book entitled *The Mountains of Oregon* and was in touch with noted contemporaries in conservation, including John Muir, Gifford Pinchot, Joseph Le Conte, and Captain Clarence Dutton, who was then working as part of the U.S. Geological Survey.

Crater Lake National Park

Steel encountered Dutton at Crater Lake on Dutton's first trip there in 1885. They discussed ways to preserve the area. From him he learned that it was critical, as a first step, that the area be withdrawn from disposal, particularly from mining and homesteading. Otherwise, the area might be lost to claimants. Dutton suggested that ten townships be withdrawn. The area's congressman at the time, Binger Hermann, also encouraged him to seek appropriate federal protection.

Steel set about to get that done. He began by circulating petitions to have Crater Lake made into a federal park. He wrote letters to most of the newspapers in Oregon asking them to circulate his petitions, and most did. He did the same with local postmasters (his brother was the postmaster in Portland). With the help of Oregon legislators, his petition found its way to the Secretary of the Interior. Steel then traveled to the nation's capital and met with him. He even managed to see President Grover Cleveland, who responded in 1886 by withdrawing the ten townships as requested. He reserved them for retention in federal hands.

Oregon Congressman Thomas Tongue then joined the cause by introducing legislation in 1898 to make Crater Lake a national park, and Oregon

**William Steel at Crater Lake
circa 1929**

Senator George McBride followed suit the following year. Oregon's legislature got behind the effort too.

However, resistance began to emerge to the legislation. Timber and grazing interests raised objections, fearing that they would lose access to resources they imagined to be there. Others in Congress worried that the designation would divert finances away from more important projects, and some argued that the state instead should care for the area. All of them combined to bottle the bills up in committee.

Steel became convinced that the key to moving the legislation forward was to adduce more evidence of the scientific importance of the lake. To do this, he worked to bring Dutton back to the lake for further research on its depth and other features. In response, Dutton came back, continuing his research for the Geological Survey. Steel also attracted other scientists such as C. Hart Merriam, as well as a visit in 1896 from the Forestry Commission of the National Academy of Science. Muir and Pinchot both came there in connection with the visit of the commission.

These visits enabled Steel not only to enlist prominent leaders in his cause but also to convince a number of them to write articles about the lake, which built further support. Steel used these articles to demonstrate the lake's importance to science and the need to protect it.

Steel also addressed the opposition from miners and farmers, pointing out that there were no resources valuable for these purposes in the Crater Lake Reserve. Steel then got the new congressman from Oregon to sponsor the bill. But still the legislation sat waiting as other bills to establish new national parks moved along to enactment, including that for the nearby Mt. Rainier National Park.

In 1901 Steel geared up his campaign once again, sending a flood of letters to editors, urging citizens to write Congress. He also sent Congress a petition representing 4000 Oregonians. This effort got the bill out of committee, but the Speaker refused to put it to a vote. Finally, Steel got President Theodore Roosevelt to prevail upon the Speaker to relent; he did this by enlisting the intervention of Gifford Pinchot, whom Steel had met at the lake at the Forestry Commission meeting. Pinchot was enthused about having a national park at Crater Lake, and he was Roosevelt's top conservation advisor.

The House then readily approved the bill, and the Senate immediately concurred. President Theodore Roosevelt signed the bill in 1902.[5]

From 1912 to 1915, Steel was actually the superintendent of the park, though he was probably less suited for this job than campaigning. No Oregonian has probably ever been a better campaigner.

---

5 For a long time, the boundaries of the park remained largely on township lines. But in 1980 Senator Mark Hatfield got legislation enacted to slightly refine the boundaries by placing them, where possible, on nearby topographic features. These boundaries had been suggested to him by Doug Scott and Larry Williams, then with the Sierra Club and the Oregon Environmental Council.

RE-VESTING THE O & C LANDS IN FEDERAL OWNERSHIP

A. W. Lafferty was not a campaigner in the mode of Steel, but he waged a relentless campaign in the courts to reclaim federal ownership of the O & C lands. He was pursuing President Theodore Roosevelt's pledge in 1903 to "clean up the O & C land fraud mess once and for all."

Lafferty was a lawyer who was sent to Oregon by the Justice Department to do this, and he was unusually tenacious and effective. Initially he was an assistant to Francis Heney, who was the lead prosecutor in the Oregon land fraud trials (more about that later). After his initial court battles, he became a congressman from Portland for two terms (1912–1915). In fact, he was elected on this issue.

Late in life, he returned to Oregon to resume the struggle. He even approached me then about picking up where he left off (I declined; I had other priorities). He was then gaunt and stooped, but his eyes blazed with intensity.

Over two million acres of richly forested lands in the hills of western Oregon are now largely administered by the Interior Department's Bureau of Land Management. Lafferty played a pivotal role in getting them back.

They had been granted under an 1866 law to the Oregon and California Railroad company[6] under a promise that it would build a railroad in the two states from Portland to Davis in the Sacramento valley. The line to the California border was supposed to have been completed by 1875, but it was not until 1887. Only 197 miles in both states were built by 1875, as the railroad companies struggled with financial difficulties.

As was customary, the railroad was granted mainly alternate sections of lands in a strip of land on ten miles on either side of the route of the new line. These were commonly called "checkerboard" lands. They got the grant lands as construction progressed. Once patented, they could be sold and some were, often under legally dubious circumstances; some believe that three-fourths of these transactions were illegal. These dubious conveyances were often overlooked for various reasons, including pressure from

---

6 In Oregon incorporated under the name of the Oregon Central Railroad Co., and later known as the "East Side Company."

Alsea Falls on O & C lands

Oregon Senator John Mitchell, who had once been the attorney for the O & C Company. Lafferty asked by what right that firm, and those who were successors in title, should be able to keep this land. And in 1869, Congress imposed additional conditions, including a strict requirement that the grant lands only be sold to "actual settlers," which was seldom the case. In the same year, a similar arrangement was made to induce the construction of a wagon road from Coos Bay to Roseburg. In this case, the builders would get three miles on either side of the road. It ran afoul of the same provisions.

In 1907 Oregon's legislature sent a memorial to Congress asking that the federal government enforce the terms of the original grant, or get the grant lands back. Many of them had long been in limbo as unpatented grants on which no local taxes were paid. While timber companies showed some interest in acquiring these lands, few of them were suited to agriculture by farmers who could be actual settlers.

Congress responded in 1908 by authorizing an enforcement lawsuit, which Lafferty brought in September of that year, to reclaim these lands. The litigation slowly moved through the federal courts, bogging down at times as many interested parties intervened and filed countersuits. Some of the confusion was cleared away in 1912 when Congress passed legislation to grant clear

CHAPTER 1—The Foundational Federal Reserves

title to those who were styled "innocent purchasers." They got 371,000 acres in this fashion.

In 1913, the district court judge in Portland (Charles E. Wolverton, who was hearing the case) ruled that the O & C Company should forfeit its unsold lands because of failure subsequently to fulfill the various required conditions. But when it got to the U.S. Supreme Court on appeal in 1915, the Supreme Court muddied the picture even more by its ruling, after reflecting on seventeen volumes of record. It ruled instead that the company should be enjoined from taking any further actions that would violate the conditions which it characterized as "enforceable covenants," not as "conditions subsequent." It enjoined any further disposition of these lands, or timber sales on them, until Congress had a reasonable opportunity to decide how the disposition of these lands should finally be handled; otherwise, the railroad company would have to comply with the original terms of the grant.

Each aspect of the court decision brought more arguments and counter-claims because these lands were worth tens of millions of dollars, with as many as 15,000 claims for them filed mostly by speculators.

In 1916, Congress had to put an end to the dispute and finally decide the disposition of the lands. Senator George Chamberlain had campaigned for years to reopen this matter, particularly when he had been Oregon's governor. Now he led the way in the enactment of the Chamberlain-Ferris Act, in which Congress decided to re-vest these lands in the hands of the U.S. government, specifically the Interior Department, while paying some of the counter-claims of the railroad that had succeeded to the title. At this point, there were 2,891,000 acres left to be re-vested. This Act also recognized that most of them were timberlands and not suited to farming, directing that the timber be sold by competitive bids "as rapidly as reasonable prices can be secured…in a normal market." Cut-over timberlands were then to be sold.

In the 1950s, Lafferty brought new suits to bring even more of these lands under the formula Congress had set for distributing timber receipts from these lands. Once again, he prevailed before the Supreme Court.

If Lafferty had not been so tenacious, these lands might not be in federal hands today. He deserves to be remembered.

The fate of the Coos-Bay Wagon Road lands was also settled by Congress a few years later (1919) when Congress decided to also re-vest 93,000 acres of them in the same manner as the O & C lands and directed them to be managed as the Chamberlain-Ferris Act provided.

Even though the Chamberlain-Ferris Act brought order out of chaos, it was not welcomed by all interests. Commercial interests were unhappy because they were deprived of the opportunity to acquire this timber at "bargain-basement" prices, and the final private owner of these timber lands, the Southern Pacific Co., was outraged at losing these valuable forest lands. And the Forest Service was out-of-sorts because it felt that these lands ought to have been turned over to it for management.

In the ensuing decades, the Interior Department and the Forest Service found much to argue about with regard to the O & C lands. The Forest Service felt that they were not being managed in an efficient and professional manner. And it argued about who should get what came to be called "the controverted lands," which were 472,000 acres of O & C lands that were intermingled within the boundaries of its national forests. For many years, the Forest Service sold the timber within them, later putting the receipts in a suspense account to be distributed once the issue was settled. It argued that their final disposition had never been settled and that they had long been managing them.

To tamp down the level of this criticism, Senator Charles McNary secured an amendment to the Chamberlain-Ferris Act in 1928 authorizing the GLO to adopt rules and regulations to guide its timbering operations, though the Forest Service contended they did too little to implement them.

The most damaging criticism was that the Interior Department was just liquidating the timber stands instead of reforesting them and managing them for the long run. It also provided little oversight of timber sales, but GLO argued that the Chamberlain-Ferris Act told them to do this and did not contemplate that the lands would be kept permanently in federal ownership.

Western Oregon counties preferred having these lands under the Interior Department because they ultimately got a larger share of the receipts from timber sales: seventy-five and later fifty percent in contrast to the twenty-five

Three Arch Rocks on Oregon Coast

percent from Forest Service sales; this led to later problems when the sale level was reduced. They had become dependent on this extraordinary level of largesse. And they felt that the Interior Department was inclined to log more.

Finally, in 1936 the Interior Department saw the need for more statutory guidance on managing these lands. Under the leadership of Secretary Harold Ickes, the Interior Department at that time saw itself as a "Department of Conservation" and did not want to be seen as wanting in its stewardship. He pulled together a team that drafted remedial legislation, which passed in 1937 and is now seen as the organic act for administration of the O & C lands.

This Act specified that these lands would be retained and stay under the management of the Interior Department, under a new management unit (leaving the GLO). They were to be managed under the principle of sustained yield to provide a permanent supply of timber so as to contribute to the economic stability of local communities and industries. But this was not the only focus for their management: they were also to be managed to provide for recreation and to regulate stream flow and protect watersheds—all in a manner that later came to be known as multiple use.

In a provision that still excites controversy, the Interior Department was directed to see that the "full allowable cut...be sold annually, or so much thereof as can be sold at reasonable prices in a normal market." Arguments continue over whether this mandates maximizing sustained-yield cuts or whether it allows balancing various multiple uses in determining the level of the allowable cut.

Following this enactment, the Interior Department put together a new organization to manage these lands, reclassifying its lands and conducting inventories, with professional foresters being hired. Selective logging was practiced (for a while), and the Civilian Conservation Corps (CCC) used to reforest cut-over lands. In 1946, President Truman set up the BLM to manage these lands, as well range lands elsewhere used for grazing.

The 1937 organic act included a provision that authorized the managers to form what were called Cooperative Sustained Yield Units with firms that had large tracts of timber. This ostensibly allowed balancing cutting operations over a larger area, but recreationists suspected it rationalized abusive levels of logging on widespread acreages. When efforts were made in the late 1940s to form them, that was not done because small firms (not having their own tracts of timber) were disadvantaged and multiple use was undermined. Among others, the state Izaak Walton League opposed their formation.

The O & C lands were finally out of their formative stage when timber was all that mattered.

NATIONAL WILDLIFE REFUGES

Three Arch Rocks

The oldest wildlife refuge in the West is among the rocks off the shore of the Oregon coast near Tillamook. It started as the Three Arch Rocks and now is part of a larger refuge along much of the coast.[7]

7 This refuge is now comprised of offshore rocks, headlands, and estuaries. It provides habitat for a total of 1.2 million seabirds and thirteen species on 1853 rocks. It consists of six units: Three Arch Rocks, the Oregon Islands, Cape Meares, Nestucca Bay, Siletz Bay, and Bandon Marsh. It includes all sorts of special features: wintering ground for the entire world population of the Semidi Islands Aleutian Cackling Goose (at Nestucca Bay), Oregon's largest Sitka Spruce (at Cape Meares), and the southernmost coastal sphagnum moss (at Neskowin marsh in the Nestucca Bay unit).

William L. Finley
circa 1908

While it consists of merely fifteen acres, it is notable for the amount of habitat it provides. Being on offshore rocks, it cannot be reached by many predators. It harbors Oregon's largest colony of Tufted Puffins and the largest breeding colony of Common Murres south of Alaska. As many as 230,000 seabirds now occupy these rocks, including twelve species.

Two Oregonians persuaded President Theodore Roosevelt to use administrative authorities in 1907 to set aside the refuge. They were William Finley and Herman Bohlman, who were both expert naturalists and committed birders. Since about 1900, they had been observing the birds on these rocks and began to notice them being harassed.

They decided they needed to document the damage being done. So in 1903, they photographed the birds there, documenting their problems. One

Sunday, they actually witnessed hunters collecting eggs and shooting birds there, as well as sea lions. The eggs were shipped to restaurants in San Francisco. They saw thousands of birds being wiped out, which appalled them. They resolved to do something, actually managing to row a boat out to the rock to establish a camp where they took thousands of photographs to document the damage.

Afterwards they first tried to get the state to make it a sanctuary. While they could not, they did obtain a Model State Bird Law that they used to block shooters who sought to travel out to the rock. In the process, they drew the derision of many political figures, but they were not daunted.

When President Roosevelt visited Portland briefly in 1903, they approached him to appeal for help. He showed interest but did not have the time to examine their photographs; he invited them to come to see him later in the White House. The next year—following the election—they made that trip and were able to show him their documenting shots. He was convinced, but first explained that he needed to get a new law making trespass in federal refuges a crime, as well as harming birds, their nests and eggs; violators could then be arrested. It took him a few years to get this legislation.

By 1907, he had what he needed, proclaiming this refuge, and thereupon began the process of reserving such areas elsewhere in Oregon and throughout the West.

President Roosevelt considered himself something of a naturalist and cultivated continuing relationships with experts on the habitat needed by birds in various parts of the country. William Finley (1876–1953) became his contact in Oregon. Becoming known not only as an expert birder and naturalist, he was a writer and photographer as well. As a leader in what is now known as the Portland Audubon Society,[8] he and they were trying to stem the decimation of waterfowl to get plumes for the millinery trade. In due course, Finley became the head of the Oregon Fish and Wildlife Commission. He also was instrumental in establishing other refuges.

---

8   Known then as the Oregon Audubon Society.

Wildlife Refuges East of the Cascades

The marshes of eastern Oregon were the state's most important waterfowl habitat. They provided key stops on the Pacific flyway, supporting immense numbers of birds. In 1905, Finley and Bohlman visited the Klamath marshes and saw their incredible value. These marshes were then not only threatened by plume hunters, but tons of the ducks were being shipped to California for restaurants. The two began filming the threats, and even arranged for a patrol boat to discourage plume hunters.

They were also quite familiar with the value of the Malheur marshes further east. On their visit to Washington, D.C., in 1906 to see President Roosevelt about the threats to Three Arch Rocks, they also made a plea for action to save the Klamath and Malheur marshes. Now that Roosevelt had his trespass law, he acted with alacrity. 1908, he set aside 108,000 acres as the Malheur Lake Refuge (its name has now changed). It was then particularly important in providing habitat for the Great Egrets, whose plumes were avidly sought for the millinery trade. Today as many as 320 species of birds are counted there, and its size has grown. He also set aside a Klamath Refuge of 81,000 acres.

However, both were burdened by nearly fatal problems. Until the 1930s, the Malheur refuge was deprived of necessary water. Ranchers had erected dams on the Blitzen River that were diverting water from its lakes. Droughts then also aggravated the problem. Finley worked with various heads of the Biological Survey (predecessor of the Fish and Wildlife Service) to acquire the dams and bring another 65,000 acres into the refuge.

The Klamath Refuge (spanning the state line) had been declared within a drainage project of the Bureau of Reclamation, dating from 1905. Its marshes were being drained for farms and its water used to irrigate them. Finley denounced this drainage as a "crime against our children." But despite protests, it went inexorably on.

At the time, Roosevelt did not see the conflict between reclamation and refuge status. He was only thinking of bringing an end to the depredations of the egg and plume takers, which this designation did accomplish. He was not focusing on the need to preserve the habitat for the birds. In fact, he repeatedly used this approach in the West by overlaying refuges on drainage projects (he did it seventeen more times); it never worked. His conservation

Malheur National Wildlife Refuge

Upper Klamath National Wildlife Refuge

CHAPTER 1—The Foundational Federal Reserves

advisor, Gifford Pinchot, did not want to point out this conflict because he was an inveterate reclamation booster.

In 1977 Congress at last clarified that waterfowl management was the primary purpose of this refuge, with irrigation supposed to be a secondary use that was only permitted to the extent it was compatible. But notwithstanding, the conflict continued.

In recent decades, the administering agency has patiently assembled parcels to build back portions of the refuge around Upper Klamath Marsh. Fifteen thousand acres were put in it when the Klamath Indian Reservation was terminated. Now there are 40,000 acres in it, comprised of some marshes, reclaimed marshes, and adjoining meadows. In 1998 federal judge Michael Hogan ruled that the Indians there had water rights there that were superior to the irrigators, giving them leverage, it is hoped, in restoring the marshes. And the Oregon Water Resources Department has now found that their water rights are superior to those of all other claimants (by being first in time).

But problems of water quality continue because the reclamation project's farmers have not given up; they continue to compete for water. And the locality still wants water to be diverted for power.[9] Nonetheless, birds flock here, including Greater Sandhill Cranes. Half of the Yellow Rail in the western United States breed here.

Despite its problems, this refuge provides valuable habitat. Finley is celebrated as the father of Oregon's waterfowl refuges.

Incidentally, 4800 acres near the Klamath refuges have now been protected as a roosting area for Bald Eagles: the Bear Valley National Wildlife Refuge. Located a few miles north of the state line and just west of Highway 97, this

---

9 Many have been led to believe that these problems have been resolved through a settlement negotiated with stakeholders reached in 2009. However, the Interior Secretary had to find this agreement to be in the public interest by March of 2012, and did not do so, and Congress had to approve the arrangement, which it has not yet done. Moreover, a number of environmental groups did not join the agreement, and some believe it is deeply flawed, giving too much control to the irrigators and the power company (which wishes to keep the dams running as long as possible). Others doubt the required money will ever materialize and feel the supporting science is wanting. Now the Klamath County commissioners have withdrawn from it.

forested area was given protection in 1988. As many as 500 of these eagles flock here in the winter to feed on ducks in the Klamath refuges. The hillside old growth forests facing to the northeast provide ideal habitat for them. It is not open to the public.

And just a couple dozen miles to the northwest on Pelican Bay, railroad tycoon E. H. (Ned) Harriman had a lodge he called Harriman Springs, where he spent summer time as he aged. In the context of conservation, it should be noted he was a friend and supporter of John Muir.

Hart Mountain Antelope Refuge[10]

This eastern Oregon refuge came about in a different fashion. It aimed at protecting upland habitat for the Pronghorn Antelope. It was not the product of activists. Instead, it was the product of a combined effort by agency wildlife experts and homegrown tourism boosters. One can infer that they used back-channel contacts among VIPs to produce the refuge in record time in the 1930s.

Since the 1920s, the Oregon Game Commission had been working quietly with the federal Biological Survey (then in the Agriculture Department) to create this refuge. At that time, the regional director of that agency in Portland was Ira Gabrielson, who later became the first director of the Fish and Wildlife Service. He was quite intent on expanding the refuge system. Undoubtedly, the two agencies worked together to pursue the establishment of this refuge for the Pronghorn Antelope.

At that time, populations of the pronghorn had been severely reduced by heavy hunting. The pronghorn is the fastest land mammal in the western hemisphere, and worldwide second only to the cheetah, and capable of sustained speeds longer than the cheetah. In Oregon, only a remnant of the population remained around Hart Mountain, and even it was in trouble. Decimation of the pronghorn there was so bad that the state Game Commission had sued to stop hunting there.

---

10 For many years, this area was called a "range," but now it is called a refuge. The Fish and Wildlife Service used to label its upland reserves as ranges, and its waterfowl reserves around marshes as refuges.

Antelope on Range in Eastern Oregon

The Game Commission wanted local support for this project. It turned out that there were officials of the Chamber of Commerce in nearby Lakeview who also were disturbed by the harm being done to the antelope. In the midst of the Depression of the 1930s, they saw the antelope as a potential feature to attract tourists and boost the local economy. They organized a group to bring important people from throughout Oregon to Hart Mountain to learn about the antelope and its problems. They called it the Order of the Antelope. Its first chairman in 1931 was Marshall Dana of Portland of the Game Commission, who was also the editor of the *Oregon Journal.*

Dana understood the importance of building public support, and he knew fellow newspapers editors around the state. He knew that if the editors went to the trouble of attending one of their affairs, they would probably write supportive editorials. So he had the Order invite many of them to annual summer gatherings on the summit of Hart Mountain to enjoy themselves and hear their story. On their first trip, he made sure that Gabrielson attended.

They continued to hold these affairs year after year—enlisting more and more support around the state. Exactly how this support and that of the wildlife agencies translated into successful action is still not exactly clear. But many of the key people outside of the state knew each other because they came from the state of Iowa. These included not only Gabrielson, but others too. At that time, the head of the Biological Survey was "Ding" Darling who came from Iowa, as did the Secretary of Agriculture, Henry A. Wallace, Jr. It can be presumed that Gabrielson urged Darling to take the matter to Wallace. And for that matter, Darling knew Wallace from their Iowa days.

A person who had been friends with Wallace's father was then Oregon Senator Charles McNary. It would have been logical for Dana to have been in touch with McNary to get Wallace to move the proposal along to President Franklin Roosevelt (remembering too that Wallace had influence with FDR and was later his Vice President).

At any rate, in 1936 President Franklin Roosevelt issued an executive order reserving this refuge out of the public domain as the Hart Mountain Antelope Range, as well as its counterpart twenty miles south in Nevada—the Sheldon Range. Hart Mountain provides summer range for the antelope, while Sheldon provides winter range. As a result of steady additions, today 278,000 acres are protected in the Hart refuge.

But the lands there were not well protected for a long time. When it was established, the number of cows grazing on the refuge was reduced from 10,000 to 4,000 and numbers of sheep were reduced too, but over time these remaining numbers had a disastrous impact on the habitat. Most of the characteristic vegetation disappeared, and many of the streams either disappeared or became entrenched; most of the springs dried up. Eventually, a lawsuit brought by environmentalists (in 1991) induced the Fish and Wildlife Service to suspend grazing there for fifteen years. Now the vegetation is coming back, and antelope numbers are at record levels (1900 of them). The habitat now supports over 300 species of wildlife, and it is a fine example of native ecosystems in this high desert environment. Because it supports the imperiled Greater Sage Grouse, it has been declared one of Oregon's three Important Bird Areas (along with the nearby Trout Mountains).

One can hope that this condition will continue.

CHAPTER 1—The Foundational Federal Reserves

REFERENCES

Leonard Rakestraw, *A History of Forest Conservation in the Pacific Northwest: 1891–1913* (New York, Abro Press, 1979); 1955 Ph.D. dissertation, University of Washington.

Bobbie Snead, *Judge John B. Waldo, Oregon's John Muir* (Bend, Maverick Publications, 2006).

Gerald W. Williams, *Judge B. Waldo: Letters and Journals from the High Cascades of Oregon 1877–1907* (Roseburg and Eugene, USDA Forest Service, 1992).

Gerald W. Williams and Stephen R. Mark (eds.), *Establishing and Defending the Cascade Range Forest Reserve from 1885 to 1912* (Portland, USDA Forest Service, 1995).

Ronald Eber, "John Muir and the Pioneer Conservationists of the Pacific Northwest," in *John Muir in Historical Perspective,* Sally Miller, ed. (New York, Peter Lang, 1999).

Rick Harmon, *Crater Lake National Park: A History* (Corvallis, Oregon State University Press, 2002).

Erik Weiselberg, "He All But Made the Mountains: William Gladstone Steel, Mountain Climbing, and the Establishment of Crater Lake National Park," the *Oregon Historical Quarterly* (spring 2002), p. 50.

Steven R. Mark, "Seventeen Years to Success—John Muir, William Gladstone Steel, and the Creation of Yosemite and Crater Lake National Parks" (*Mazama* 1990), vol. 5, p. 12.

Bureau of Government Research and Service, "The O & C Lands," University of Oregon and O & C Counties, 1981 [special report].

David M. Ellis, "The Oregon and California Railroad Land Grant, 1866–1945," *Pacific Northwest Quarterly*, vol. XXXIX (October 1948).

Joseph S. Miller, *Saving Oregon's Golden Goose: Political Drama on the O & C Lands* (Portland, Inkwater Press, 2006).

Douglas Brinkley, *The Wilderness Warrior: Theodore Roosevelt and the Crusade for America* (New York, Harper Collins, 2009), chapters 19 and 24.

William G. Robbins, *Landscapes of Conflict: the Oregon Story, 1940–2000* (Seattle, University of Washington Press, 2004), chapter 4.

William G. Robbins, *The Early Conservation Movement in Oregon, 1890–1910* (Oregon State University, 1975).

Worth Mathewson, William L. Finley: *Pioneer Wildlife Photographer* (OSU Press, Corvallis, 1986).

Hallie Huntington, *History of the Order of the Antelope* (Lakeview, Lake County Examiner Press, 1969).

Ira Gabrielson, "Origin of the Ancient Order of the Antelope," *Guano Creek Morning Bloomer* (Lake County, 1960).

CHAPTER 2
THE STATE AND OTHERS DO THEIR PART

Protecting Oregon's Beaches

Sometimes the pioneering action was taken by a high official who identified with the public interest. In this case, it was the governor.

Oswald West, the governor in 1913, was a progressive Democrat who owned a summer home at Cannon Beach. On a horseback trip that year along the beach from his home south to Nehalem, he reflected on the value of the beach, which was being used by wagons carrying freight and the mails. Beginning in 1899, the legislature had declared the first thirty miles south of the Columbia River as a highway for such use.

But the beach south of that point could be closed or lost. In fact, by then some twenty miles had been lost in connection with the sale of state-owned tidelands. "We have got to put a stop to this," West insisted. "Let's not sell any more tideland. It's too valuable."

To achieve that end, he sought and got legislation declaring the beaches of Oregon to be public highways, making it more difficult to sell public tidelands and to assure the public of access to the state's beaches. "No local, selfish interest should be permitted…to destroy…this great birthright of our people," he proclaimed.

For this action, he has become an iconic figure in Oregon, with a classic coastal state park named after him near where he had his epiphany—the Oswald West State Park.

Governor Oswald West Campaigning at Newport in 1912

Alas, in the mid-1960s, it was discovered that the state only owned the wet sands, and not the adjoining dry sands. A great new controversy then erupted over this issue, but West had planted the seeds that eventually kept the beaches public. In later chapters, that story will be told.

Oregon's State Park System

Oregon's state park system began to take serious form in the 1920s. While a few scattered parks had been donated to the state at an earlier time, it really gathered momentum in the 1920s as an outgrowth of outcries over logging along well-traveled state highways.

In this case, it was then governor Ben Olcott (1919–1923), who became incensed over the destruction of forests by logging along the road between Cannon Beach and Seaside. Though a Republican, he was a protégé and friend of Os West, who had appointed him as Secretary of State. He ascended to office when Governor Withycombe died after two months as governor.

To combat the spread of this practice, Olcott appointed citizen committees to study the challenge and sought public support to stop it, thereby becoming the leader of Oregon's scenic preservation movement. He viewed the state's forests as the "crowning glories" of its attractions. In a special message to the legislature in 1921, he called for legislation to protect "virgin stretches of forest along highways…to leave the beauties of the landscapes unimpaired."

In that year, he persuaded the legislature to outlaw destructive tree cutting along state highways and to authorize the state to acquire rights-of-way along those highways for the purpose of preserving scenic beauty. The state Highway Commission was authorized to acquire rights of way within 300 feet of the center of the highways, and soon began to acquire roadside parks and waysides. The twelve-mile-long Van Duzer Forest Corridor on Highway 18 to the coast is probably the best embodiment of what Olcott sought.[11]

Olcott's efforts bore further fruit in 1925 when the legislature authorized the state to acquire tracts away from state highways (no longer having to be within the 300-foot zone) for the purpose of tree culture and the provision of parks and recreation grounds.

But these pioneering provisions were only fully used when Sam Boardman was hired to head the program. As the first superintendent of Oregon's State Park System, he, more than anyone else, was responsible for building the system. He provided the vision that guided the development of the system, even though, ostensibly, he was just a bureaucrat. While he worked with others to put 60,000 acres into the system, he provided the vision and drive at the outset and continued in that role for over two decades (from 1929 to 1950).

At the outset, Robert Sawyer of Bend collaborated with him.[12] Sawyer was a member of the state Highway Commission in the late 1920s, and it was he who persuaded the commission to hire an official to supervise the parks agency that was then under the highways commission. They hired Boardman. And in the 1930s, Boardman also collaborated closely with Jesse M. Honeyman of the Oregon Roadside Council, which pushed hard for more

---

11 Sam Boardman set it up in the 1930s.

12 Sawyer was also the editor then of the *Bend Bulletin*.

state parks. Sometimes Boardman got Honeyman to prod commissioners to approve his projects. Other groups, such as Save-the-Myrtlewoods, helped acquire groves for state parks in southwestern Oregon.

When the highway commission elevated Boardman into the position, "they had no idea of what they were getting," says park historian Thomas Cox. When Boardman had been one of their engineers from eastern Oregon, he was known for assiduously planting trees along those highways. When he came to be in charge of state parks, he continued to be concerned with having forests along the highways in western Oregon as well. He bought cut-over land along highways when it was cheap—looking ahead to when the forests would grow back.

He was determined to keep state parks as natural as possible. For instance, he negotiated with superiors in the Highway Department to change the alignment of Highway 101 south of Arch Cape to spare choice old growth forests in what is now Oswald West State Park. He opposed developments that did not belong in the parks; fought to keep out non-native plants; and sought new lands for the parks far in advance of demonstrable need.

He sought donations from the affluent; talked owners into reducing their asking price for land; got options; bought lands on credit; and even reached into his own pocket—all so that the system could grow.

He regarded the Oregon coast as a special place and acquired as much land there as possible between the highway and the beach. In the 1930s, he saw a fleeting opportunity to buy land there when the price was low. Now there are over eighty state park system units there, with Cape Lookout Park one of the best (2014 acres).

While at first he focused on trees and scenery, in time he also saw the importance of protecting habitat for wildlife and lands of scientific and historic importance.

By being associated for many years with the highway commission, the state park system shared in the receipts from gasoline taxes on highway users and benefited, but after a while highway commissioners began to get restive with Boardman's free-wheeling approach. They did not always understand why he was acquiring parks far from highways as they tended to see state parks as merely stopping points along highways. As Boardman's vision broadened, he had greater troubles in bringing the commissioners along with him. And his ally, Sawyer, had long left the commission.

But Boardman lasted a long time and laid the foundations for a park system for the state that meets professional standards. Finally, in 1989 in response to a proposal from Governor Neil Goldschmidt, it was at last removed from the Highway Department and made an independent unit, where it no longer received gas tax receipts. After a few years of being dependent on the vagaries of funding from the legislature, Brian Booth led the campaign to earmark a portion of the state's new lottery proceeds for state parks. Booth was the first chairman of the new Oregon Parks and Recreation Commission. When the state's voters mandated that fifteen percent of the lottery proceeds would go these parks (it was also used to protect wildlife and water), the financial stability of the state parks program was finally assured—though this did not guarantee any increase in funding. The voters have continued to stand by this allocation, even in times of budget crunches.

Probably its most notable acquisition, Silver Falls State Park in the foothills east of Salem started as a small state park in 1931. Resorts began to be built around the area in the 1880s. June Drake, a photographer in Silverton, for decades championed its establishment. But today it includes 8700 acres along the north and south forks of Silver Creek and is Oregon's largest state park. It features ten significant waterfalls, with the tallest at 177 ft.

It is so spectacular that in the 1920s and '30s the area was studied as a possible national park. When the National Park Service (NPS) got involved in the 1930s, it rejected the idea of making it a national park because of the amount of damage and development. The area had been burdened by logging, farming, and resorts; the Park Service even noted that logging had reduced the volume of water flowing over the waterfalls. But this suggested that it ought to be reclaimed, and they instead turned it into what they then called a Recreation Demonstration Area. It was one of only two in the West. They then acquired about 6000 acres there. In reclaiming this land, they planted trees and removed evidence of past resorts and farming—building

South Falls, Silver Falls State Park

North Falls, Silver Falls State Park

Chapter 2—The State and Others Do Their Part

Smith Rock State Park

trails instead. Through the CCC, a beautiful new stone lodge was built near South Falls, as well as other improvements.[13] The furnishings of the lodge were designed by Margery Hoffman Smith, who impressed Boardman with her design work at Timberline Lodge.

In the late 1940s, the NPS turned the area over to the state for use as a state park, becoming the crown jewel of the state's park system. For the most part, it is a legacy of the New Deal.

Other spectacular state parks exist too. One of those, also a gem, is Smith Rock State Park near Redmond. While it is only 651 acres, it features some of the country's best climbing rocks. Comprising tuff and basalt cliffs, the rocks wind along the Crooked River. It is one of the most photographed sites in Oregon.

The land was acquired through gifts and purchases in the 1960s and '70s, being made a state park in the 1990s. When a resort threatened the site, farmers, 1000 Friends of Oregon and the state Sierra Club stopped it.

---

13 Over the time of the Depression, the CCC installed facilities in 45 state parks in Oregon; they did this under the supervision of the National Park Service.

Portland Does Something Important: Forest Park

Portland boasts the largest forested city park in the country: Forest Park in the hills northwest of downtown. This is a nature park of over 5000 acres; access is via an extensive system of trails.

Some of the nation's foremost park planners promoted it: John C. Olmsted (in 1903–1907), E. H. Bennett (in 1912), and Portland's own expert park director, Emmanuel Mische (1908–1915). Even Robert Moses, the once-vaunted city planner, urged it (1943).

But it almost did not come about. Voters turned down bond measures to support parks there and elsewhere, wood cutting began, and developers were given free rein. They scooped roads into the hills to provide access to planned subdivisions. But as a result of bad planning and bad times, no homes were ever built, and washouts followed on the steep slopes. As the project became moribund, the county took over the tax delinquent lands.

In the 1920s and '30s, it turned into a "no man's land." Alder grew up in the many spots ravaged by tree clearing and fires. But as trees grew back, it began to be used by Boy Scouts and outdoor clubs for hikes since it was the closest forested open space. In their spare time, people such as the Forest Service's regional recreation director, Fred Cleator, began to plant trees, and new forests thrived. Outdoor clubs such as the Mazamas and the Trails Club of Oregon joined in.

In the process, a local businessman who was active in the Trails Club saw new potential. He was Garnett "Ding" Cannon, then president of the Standard Insurance Company and of the Trails Club. In a few years, he also became the president of the Federation of Western Outdoor Clubs (FWOC). In 1944, he persuaded the City Club of Portland to do a study of what should become of the area. A committee of five citizens was appointed, with Cannon chairing it. It took a close look at the area on the ground, and a year later endorsed making the area a wild city park. But then interest suddenly ebbed as interest emerged in exploring in the area for oil (but to no effect).

But Cannon knew that something had to be done to regain the initiative. He used his position with the Federation of Western Outdoor Clubs to call a meeting to develop a plan of action. Held in 1946 at the office of the Mazamas, it was attended by thirty-four people. They made Thornton Munger their chairman and Cannon their vice chairman. Munger had recently retired as chief of the Forest Service's regional research office. He was enthusiastic about the idea and shouldered the burden.

They decided to enlist as many as possible in this effort, sending letters soliciting support to such groups as the PTA, garden clubs, angling groups, Audubon Societies, Boy Scouts and Camp Fire Girls, as well as outdoor clubs. They even reached out to business and labor groups. They styled themselves as the Group of Fifty (later becoming the Friends of Forest Park).

In the process, they managed to enlist the backing of the city Planning Commission and proceeded to petition the City Council to acquire these lands and make them a park. As a result of these organizing efforts, public support solidified behind the proposal. On June 4, 1947, the City Council voted unanimously to adopt the plan and to move forward. The park became a reality in 1948 when it was dedicated.[14]

Cannon and Munger gave us all a lesson in how to resurrect an idea that would not die because it was so good.

Thwarting Abuses and Mistakes

Land Frauds

In the 1890s and early 1900s, immense amounts of Oregon's prime timberland found their way through fraud into private hands. Through various ruses, millions of acres were looted from the public domain. Among those defrauded were the federal government, the state government, and Indian tribes such as the Siletz.

---

14 Tryon Creek State Park in southwest Portland serves many of the same purposes, but on a much smaller scale. It was established in 1975 as the result of citizen activity, principally by Lucille Beck.

The federal government lost the most: 3.8 million acres. Under its Timber and Stone Act, 160-acre plots were disposed of to prospective settlers, but they were supposed to actually live on the land. Few did; many had not even visited the parcel they claimed. These "dummy claimants" were often rounded up at saloons in towns and taken to land offices to falsely swear to their intent. Then they quickly signed over title to speculators and large timber combines in return for nominal payments. The Booth-Kelly Company got 90,000 acres of its forestland in this fashion and more through other scams.

When new forest reserves were in the offing, the ringleaders bribed officials of the General Land Office to obtain advance information to enable dummy claimants to file claims for prime timberland. They would later exchange them for isolated stands that timber companies would seek. To establish these claims to these inholdings, sometimes they would falsify the dates on their claims. Government surveys were even burdened by irregularities that they used to their advantage.

Railroads were also involved in fraud, getting checkerboard land for railroads they never built and selling these lands at prohibited prices. The problems of the O & C claims have already been described. But the Northern Pacific also was involved. Even though most of its trackage was in other states, it obtained 320,000 acres of prime timberland in Oregon through exchanging its less valuable acreage in Mt. Rainier National Park for these much more valuable commercial forest lands.

The state of Oregon also suffered from fraud. When it was admitted as a state, it was given 4.3 million acres of largely timbered lands to support its schools. Instead of holding onto these lands and managing them prudently over time, our legislature approved selling them off at once to settlers for $1.25 an acre. But few of those who got these forested lands were actual settlers. They were primarily timber companies.

For a period in the early 1900s, Oswald West was the State Land Agent who successfully recovered 900,000 acres that were obtained from the state through fraudulent representations. Through this service, he gained a reputation as a reformer. He warned against "land pirates…whose sole desire… is to loot the public domain."

But today, Oregon's school lands (1.6 million acres) are mainly desert lands in Harney and Malheur counties. Forest lands in Elliott State Forest near Reedsport are an exception. Most of the forested school lands have vanished.

When President Theodore Roosevelt's Secretary of the Interior, Ethan Hitchcock, learned of possible land frauds and cover-ups by his local officials, he had secret service agents investigate. They found amazing evidence. Soon the Attorney General appointed a veteran prosecutor, Francis Heney, to pursue cases against everyone involved. One of his assistants was A. W. Lafferty, who specialized in the O & C cases.

The web of fraud was broader than anyone imagined. It involved much of the Oregon's Republican elite, including its longtime U.S. Senator, John H. Mitchell, Congressman John Williamson, and the U.S. Attorney for Oregon, John Hicklin Hall. Hall was charged with failing to prosecute fraud cases. Also implicated was the local head of the General Land Office, Binger Hermann. Hermann had led investors to lands they would like to obtain, even when they were not qualified.

At the time, some thought that the ringleader was Stephen A. D. Puter. It turned out that he had been recruited for this role by a railroad official, Ned Harriman. The scheme all came out when Puter testified against Harriman in retribution for being fired by Harriman. Eventually Puter even wrote an exposé in jail. And other members of the conspiracy then decided to testify for the prosecution.

As a result of grand jury deliberations, over 1000 people were indicted. Later, when Heney focused his prosecutions on the strongest cases, 126 convictions were obtained. Mitchell was convicted of falsifying evidence in trying to get one of his clients' fraudulent claims; his partners testified against him. Williamson was convicted in fraudulently obtaining grazing lands near Prineville. Hall was convicted of protecting a political supporter from prosecution. A lone jury holdout kept Hermann from being convicted. And Puter went to jail. So did most of the others who were convicted. These were known widely as the Oregon Land Fraud Trials.

Reacting to these appalling revelations, Congress transferred control of the national forests in 1905 from the Interior Department's General Land Office to the Agriculture Department and put them in the hands of its Forest Service. The Agriculture Department was then viewed as far less prone to corruption.

These convictions temporarily wounded Oregon's Republican Party, although nationally they boosted Theodore Roosevelt and his conservation crusade. The benefit, however, was short-lived, and before long President Taft pardoned Hall, and the Supreme Court overturned Williamson's conviction. Mitchell died before he went to jail.

Ever since this time, though, doubt has hung over the manner in which many of Oregon's timberlands got into private hands. But at least the era of massive land fraud was brought to an end.

Avoiding Missteps

In the late 1920s, tourist boosters in Portland pushed hard for permission to build a tramway to the top of Mt. Hood. They convinced a company (Cascade Development Company) that wanted to build a lodge there to also propose the tramway. In 1927, the chief of the Forest Service, William B. Greeley, even chaired a hearing in Portland to hear the public's thoughts. Strong testimony was heard both favoring and opposing the idea.

Greeley turned the proposal down, saying that "…we must retain substantial areas exclusively for un-motorized and non-mechanical forms of recreation." He went on to emphasize that "…we should preserve…mountain peaks having the grandeur and commanding position…of Mt. Hood."

But the company would not accept his decision. It appealed it to the Secretary of Agriculture, who appointed two different committees to advise him. The first was a committee of Portlanders, which endorsed the idea. The second did not, though it was split, 2–1. The two who opposed it were the noted park and landscape planner, Frederick Law Olmsted, Jr. and John C. Merriam, who then headed the Carnegie Institution. With this negative verdict, the idea died.

Chapter 2—The State and Others Do Their Part

A decade later (1938) a lodge was built on the south slopes of Mt. Hood, but Oregon almost did not get the rustic and charming Timberline Lodge. Portland landscape architect John Yeon, who was instrumental in persuading WPA manager Emerson Griffith to site the lodge at timberline, also tried to persuade him to choose a contemporary design for the lodge.

However, the Forest Service resisted, preferring a rustic design. It persuaded Griffith to instead choose the designer of the famed Ahwanhee Lodge in Yosemite National Park, Gilbert Stanley Underwood. Underwood produced a design that relied on local, natural materials and featured handcrafts and hand detailing.

We almost did not get the Timberline Lodge that is so beloved. Today it is a National Historic Landmark that attracts millions of visitors.

The Depression that gave us this lodge as a work relief project also killed another project—that of building a 300-mile long skyline road along the Cascades. Instead, we got a skyline trail, which was more compatible with preserving wilderness areas there.

In 1915 Lewis A. McArthur had actually proposed such a road from the Columbia River to the California line. Later he became widely known for his book on the state's place names. He foresaw various uses: access to scenery, firefighting, administration, and military use.

So in 1919, the Forest Service began to examine the proposal, sending teams into the field in the summer to scout the route. As they moved along through the forest, they put up markers for the Oregon Skyline Trail. They were distinctive diamond-shaped metal markers. They also put special blazes on the trees. The teams were led by the regional recreation director, Fred Cleator. He thought most of the road would not require more than a five percent grade. They even had the Bureau of Public Roads look at the northern part of the proposed road.

Portions of the proposed road were actually built in the late 1920s. These included the road south from the Mt. Hood region to Lost Lake, Century Drive west of Mt. Bachelor, and the road from Crescent Lake south to the Crater Lake National Park line.

But at the time the entire project was simply too costly. It was dropped during the Great Depression because the entire project was estimated to cost $1.5 million, which could not then be budgeted.

Wilderness was the beneficiary.

CHAPTER 2—The State and Others Do Their Part

REFERENCES

David G. Talbot and Lawrence C. Merriam, Jr., *Oregon's Highway Park System 1921–1989: An Administrative History* (Salem, Oregon Park and Recreation Dept., 1992).

Matt Love, *Grasping Wastrels vs. Beaches Forever* (Pacific City, Nestucca Spit Press, 2003).

Thomas R. Cox, *The Park Builders: A History of State Parks in the Pacific Northwest* (Seattle, University of Washington Press, 1988).

Marcy Cottrell Houle, *One City's Wilderness* (Corvallis, Oregon State University Press, 2010); see chapter on history.

Zeb Larson, "Silver Falls State Park and the Early Environmental Movement," *Oregon Historical Quarterly*, vol. 112, no. 1 (Spring 2011).

Thornton T. Munger, *Report on Forest Park* (1960; at Oregon Historical Society).

William Body, "How They Stole Oregon Land," in *Great and Minor Moments in Oregon History,* Dick Pintarich, ed. (Portland, New Oregon Publishers, 2003).

John Messing, "Public Lands, Politics, and the Progressives: the Oregon Land Fraud Trials, 1903–1910," *Pacific Historical Review* (Feb. 1966).

Sarah B. Monro, "Timberline Lodge: the Amazing Icon of the WPA," *Timberline* (fall 2010).

Gerald W. Williams, *The U.S. Forest Service in the Pacific Northwest: a History* (Corvallis, Oregon State University Press, 2009); chapter 4.

CHAPTER 3

THE FATE OF OREGON'S RIVERS

Cleaning Up Oregon's Rivers

In the 1920s, the Willamette River was heavily polluted. Some characterized it as an open sewer. Many other rivers in the state were no better. A forceful group at the time, the state Izaak Walton League, demanded that their problems be addressed.

A succession of agencies began to document these problems in reports they issued. In the early 1930s, a New Deal planning agency for the region proclaimed that the Willamette was the worst polluted river in the entire northwest. The City Club of Portland joined in with its own report.

In 1937, Oregon's legislature responded by passing a new law to address the problem, but the conservative governor, Charles Martin, vetoed it. It had been brought to the legislature by an aroused citizens' group—the Oregon Anti-Stream Pollution League, which was outraged.

In response, a coalition of irate conservation groups took the matter directly to the voters. Among others, it was comprised of the Izaak Walton League and the Oregon Wildlife Federation. In 1938, in the midst of the Depression, their initiative measure—the Water Purification and Prevention plan—was adopted by the people by a three to one margin.

The initiative measure set up a new agency—the Oregon State Sanitary Authority—to address these pollution problems and empowered it to set regulations to curb pollution of the state's streams by industries and municipalities. It was overseen by a seven-member board.

The agency, however, struggled to make a dent in the problem, initially avoiding legal action—trusting that encouragement would lead to voluntary compliance. It also believed that pollution could be remedied simply by dilution: i.e., by mixing pollutants with enough river water.

When a new survey found that water quality was worse than ever, the Sanitary Authority finally changed its approach. Its old tactic plainly had not worked, so it then decreed that cities would have to install plants to treat their sewage. At first cities were just required to remove solids that would settle out (called "primary treatment"). They began to appear in the early 1950s: first in Portland, Salem, and then in Eugene. Other cities in the Willamette Valley soon did the same.

But efforts to get small towns to comply did not go as well. Often voters there would not approve bond measures to build these plants. Few governors were willing to push them.

In face of a flagging effort, the U.S. Public Health Service called for more plants and for more treatment: i.e., the use of filters to remove most of the pollutants (called "secondary treatment"). As a result, in the early 1960s the Sanitary Authority began to push for more progress and specifically secondary treatment. Gradually these plants began to be built too, though Portland was a laggard.

The worst foot-draggers, however, were the pulp mills that dumped their sulfite wastes directly into the Willamette River. In the 1950s, measurements found that eighty-four percent of the pollution load came from five pulp plants—two of them in the Oregon City area and three between there and Salem. Beginning in the late 1940s, the Izaak Walton League accused them of stalling and of using the same excuses they had in the 1930s—claiming that it was too expensive to clean up. The League's spokesman, Dave Charlton of Portland, a scientist, was a meticulous but determined man. He insisted that the research program of the pulp mills was nearly a sham.

In 1951, the Sanitary Authority ordered them to cease dumping sulfite wastes into the river in the summer months, when water levels dropped and temperatures rose. Thereafter, it extended the cutoff date for ending this practice year by year. As it did this, the *Oregonian* accused the Authority of "temporizing."

When the Authority finally put its foot down in 1954, the companies claimed they would have to close down. They cried piteously that they could not comply, even though elsewhere in the country pulp mills were putting such wastes into holding lagoons or were drying and burning them.

This resistance drew further media attention, with KGW-TV producing a major documentary in 1962 entitled *Pollution in Paradise,* which Tom McCall researched and narrated when he was newsman. This exposure drew his attention both to this issue and to politics.

It was not until 1965 that the Sanitary Authority finally banned dumping by such mills throughout the year. It took this step only when the legislature gave it clear authority to close down polluting industries—if it had to.

In a few years, candidates for the governorship actually vied over promises to put an end to failures to clean up the Willamette. In 1966, both Tom McCall and Bob Straub made this an issue in their campaigns against each other. When McCall became governor, he actually put himself on the board of the Sanitary Authority to give it the backbone to finally get tough. Fortunately, in time both of them became governors of the state.

In a few years, the cleanup of the Willamette River was hailed as a great success of this period. It was the first major river in the country to have installed both primary and secondary treatment of the wastes going into it—all twenty towns and cities and six hundred factories had complied. By the early seventies, Oregon even exceeded federal requirements. In subsequent years, though, it began to decline again, but that is another story.[15]

Keeping Damaging Dams off the Main Stems of Celebrated Rivers: The McKenzie, the Umpqua, the South Santiam, and the Lower Rogue

In 1937 the Corps of Engineers proposed that seven flood control dams be built on the tributaries of the Willamette River (known as the Willamette Valley Project), including a major dam on the main stem of the McKenzie River near Vida (at Quartz Creek). It had been given broad authority in 1936

---

15 For the current status of the river, see page 216.

McKenzie River—Main Stem

to prepare plans for dams on rivers across the country to control floods and for other purposes, and specific direction to begin construction of these dams in 1938 legislation. Two of these dams were completed before World War II, with construction resuming thereafter.

The valley had been plagued with severe floods since it was first settled. By 1948, ten major floods had ravaged it, with the flood of that year giving impetus to the program for more flood control projects.

While many of the dams were generally welcomed, a dam on the main stem of the McKenzie was not. This river in the central Oregon Cascades was a prime trout stream. It was fished by many, including former President Herbert Hoover, actor Clark Gable, and Justice William O. Douglas. Anglers feared the dam would ruin the habitat for trout, as well as remove reaches of the river used by float boats. Others also feared it would flood farms and homes.

Sportsmen's groups fiercely opposed a dam at Vida, particularly the Izaak Walton League of Oregon and the National Wildlife Federation affiliate in Oregon. The opposition was led by Mert Folts of the Izaak Walton League and William Puustinen. Also involved was a group that called itself the McKenzie River Protective and Development Association. These groups orchestrated an array of opponents who pointed out the damage that such a dam would do.

Instead in 1947, they suggested that the flood control dams be put on tributaries of the McKenzie: on Blue River and on the south fork of the McKenzie near Cougar Creek. These locations would have less impact on the fishery, they argued. The opposition was so heavy that the Corps finally decided to adopt their suggestions and built the dams there instead. It is ironic now to look back and realize that these decisions eventually caused old swimming holes of mine on Blue River to be flooded.

But dam builders were not through with the McKenzie. In the mid-1950s, the Eugene Water and Electric Board (EWEB) announced it wanted a project on the upper river that would have dried up and destroyed two of three spectacular waterfalls there: Koosah and Sahalie. It had long had a small project on the lower McKenzie at Leaburg.

The local conservation community opposed the project and was joined by Senator Richard Neuberger. The opposition was spearheaded by an ad hoc group—the Save the McKenzie Association. Newspapers in both Eugene and Portland supported them. Ultimately, in 1956 Eugene voters brought the project to a halt when they turned down a bond measure to finance it. Opponents were jubilant.

But EWEB was not through and persisted. It tried again by proposing a less damaging project—the Carmen-Smith project, which was eventually approved. I joined Karl Onthank in trying to rally opposition once again, but the opponents were exhausted. And they were also divided, with the state Izaak Walton League supporting it as a more balanced proposal. The opponents had already been through two fierce campaigns to save parts of the river. This time the opposition could not stem the tide.

Down on the Umpqua River, the federal dam builders gave a little thought to building a dam there, but the editor of the Roseburg newspaper, Robert Stanton, warned them away, and the river then survived unsullied.

Conserving Oregon's Environment

Sahalie and Koosah Falls on the Upper McKenzie River

In the 1970s, the Oregon Environmental Council, joined by the Sierra Club's Oregon chapter, successfully thwarted another try. The Corps of Engineers tried to build another dam on the Umpqua at Days Creek. On that project, as well as the proposed Cascadia Dam on the south Santiam, they worked through Senator Robert Packwood to secure an evaluation by the General Accounting Office (the GAO). The GAO found both were not economically justified. After considerable campaigns, they were dropped.

Protection did not come to the lower Rogue River until 1968. At that time, there were various low dams on the middle Rogue around Medford, and struggles came later over building dams on the upper Rogue.

When Congress established the Wild and Scenic Rivers system in 1968, a stretch along the lower Rogue was one of the eight original rivers put into the system. The state legislature first tried to protect the main stem of the Rogue in 1921 when it banned dams downstream from Butte Creek in the Cascade foothills. The river was already being used by drift boats that were guiding anglers on the lower Rogue. Zane Grey celebrated it as a special place for anglers to fish, with commercial fishing have been banned in 1935.

The Wild River system was designed to preserve the free-flowing character of the rivers in it; no dams could be built. Eighty-four miles were given this protection along the lower Rogue, commencing at the mouth of the Applegate River and running downstream from there. Administration is divided between the BLM and the Forest Service.

The system is primarily a linear one, with only limited amounts of land along the sides of the rivers given this status. On the average, their boundaries include about a quarter of a mile on each side, with the full title to be acquired on only part of that. Of course, in much of the national forests the adjoining land is already public land; but private tracts (known as inholdings) are more common on BLM lands.

The formula limiting lateral boundary width and condemnation was worked out by Senator Wayne Morse, who would not agree to any open-ended approach. He seemed to be responding to constituents who owned

Lower Rogue River

lodges and cabins along the Rogue. The actual formula provides that lateral boundaries cannot include more than 320 acres per lineal mile, and that no more than 100 acres can be purchased per mile. And if more than 100 acres is already in federal ownership, condemnation cannot be used.

But this was the beginning of wild rivers designation in Oregon. In later years, Senator Mark Hatfield was instrumental in making Oregon the leading state in adding rivers to the system. At last count, it had portions of forty-nine rivers and creeks in the system, totaling eighteen hundred miles of stream length. Rivers in Oregon were added in the following years: 1968, 1984, 1988 (40 added in that year), 1994, 1996, and 2000. Only about a dozen of them are major rivers. However, about 575,000 acres altogether are protected along the banks of these rivers, which can be viewed as significant.

Some environmentalists thought Hatfield was doing this, in no small part, to placate them and offset his unwillingness to go very far in designating new wilderness areas with timber. Though many of these rivers were not threatened with large new dams, small hydro installations were always a possibility. In any event, Oregon Wild (previously Oregon Natural Resources Council) was always glad to get these designations.

The state also has had its own system of Scenic Waterways, established by a voter-enacted initiative in 1970 sponsored by then State Senator Don Willner and State Representative Stafford Hansell. After Willner's proposals along these lines had failed in two previous sessions of the legislature, this initiative passed by a two-to-one margin.

Twenty-three rivers are now on its list, as well as Waldo Lake. Its provisions, as revised in ORS 390.815, ban various activities—impoundments and landfills—and regulate diversions. It is designed to protect the natural setting and quality of the water in these streams. Since most of the jurisdiction over dams is in federal hands, the state's role is primarily relevant with regard to landfills and diversions.

The beds of navigable streams are owned by the state, but projects affecting all waters of the state, including removal of appreciable amounts of gravel, are regulated under a 1967 law. Permits must be obtained in advance, with the state consulting affected parties. Permits are less likely with regard to State Scenic Waterways and salmon habitat. The state also must be notified in advance of any development within a quarter mile of a designated river. If the state and the owner cannot come to terms, then the state has the right to buy the land involved.

The Oregon Parks Department administers the Scenic Waterways program, while the Division of State Lands administers the Remove/Fill Program.

In Oregon, the public has a common law right to float even streams that are not navigable for the purpose of recreation, such as for kayaking and fishing. And its Attorney General has ruled that this includes incidental uses under the public use doctrine, or the so-called "floatage easement." How far this extends into use of bankside areas is still to be determined.

THE COLUMBIA GORGE

The Columbia Gorge is a spectacular passage of a major river through a mountain range, with world-renowned dramatic scenery. John Muir said its

Columbia River Gorge

waterfalls were so spectacular that they deserved "...a place beside the famous falls of the Yosemite Valley." One hundred fifteen of them are found there. Nonetheless, it is the locale of never-ending tension between the forces of development and efforts to protect its scenery. It is a natural corridor for transport and communication. In places, it provides inviting places for dams and settlements. And forestry, farming, and fishing have long been practiced there. It accommodates freeways, highways, railroads, power lines, hydro projects, and towns (which want to grow).

Its first highway in 1915 reflected this tension. Its designer, Sam Lancaster, shaped a picturesque highway that trod lightly on the land, with cuts, fills, and scars kept to a minimum. He used elevated platforms and tunnels to accomplish this, as well as small-arched bridges to cross ravines. Lancaster tried hard to avoid cutting trees and went out of his way to feature views of dells, waterfalls, and vistas of the river over rustic, stone railings.

Financier Sam Hill was his partner in the project and indeed the catalyst. He held up the roadways through Switzerland's mountains and along the Rhine as models and worked with Simon Benson, Julius Meier, Governor Os West, and Multnomah County's commissioners to provide the money.

**Shepperd's Dell Bridge
circa 1916**

No one has ever been more artful than Lancaster in trying to resolve this tension. The original Columbia Gorge highway is still admired for its respect for nature. Only sections are still in use by cars, the rest now being enjoyed by walkers and bikers. Portions of it are being restored.

The park-like character of this original highway doubtless helped to induce the Forest Service, which administered much of the land there, to classify its holdings there as sort of a park. At first, a strip eighteen miles long and four miles wide was called the Columbia Gorge Park; it later characterized this area as a "Park Division." It used these designations from 1915 until the early 1970s and began building campgrounds along it in the 1920s. This was at a time nationally when the Forest Service was trying to stave off efforts to shift parts of its holdings to the new National Park Service for parks.

Though the Forest Service never pinned down the meaning of the "Park Division" designation, which was used only in the Pacific Northwest, it signaled that the Forest Service recognized that it was dealing with exceptional terrain in the case of the Gorge and Mt. Hood.

The Forest Service received even a sterner reminder of the challenge it faced in holding onto this terrain when the New Deal's Pacific Northwest Planning Commission suggested in 1937 that a park be set aside there. That proposal was given further impetus when noted Portland landscape architect John Yeon[16] backed this idea and wrote a long justification for according it greater protection.

In 1953, the Oregon legislature took a beginning step toward reducing leeway to build things in the Gorge. It established a commission for the Gorge that aimed to protect the parts of the Gorge that were not already claimed for highways, railroads, dams, power lines, and towns. The commission sought better zoning and cooperative agreements for the Gorge; it also tried to coordinate state and local efforts to protect the scenery there. For instance, Multnomah County zoned its lands there to keep out industrial development and indiscriminate commercialization. Oregon followed by classifying its shoreline in the gorge as a scenic area. The Forest Service was encouraged to make exchanges of its scattered land to block up its holdings. The state also acquired more land for a state park there. Soon Washington state followed by setting up its own commission for the Gorge.

The Portland Women's Forum ran the campaign that moved Oregon's legislature to act. Gertrude Glutsch Jensen was the leader of the Forum, and long thereafter was the chair of Oregon's commission (from 1953 to 1969). She was credited with gaining protection for the state's scenic zones (3000 acres), waging a battle to keep a steel mill out of Cascade Locks, and striving to keep commercial logging out of the forests of the Gorge. She won many awards for her work in this regard, being called by some, the "Angel of the Gorge."

But even more was to happen later, which will be related in forthcoming chapters.

---

16 Yeon was the son of the man who managed the construction on the ground of the Columbia Gorge Highway.

References

Glen D. Carter, "Pioneering Water Pollution Control in Oregon," *Oregon Historical Quarterly* (vol. 107, no. 2, summer 2006).

William G. Robbins, *Landscapes of Conflict: the Oregon Story, 1940–2000* (Seattle, University of Washington Press, 2004), chapter 8; chapter 2.

John Haubert, "An Introduction to Wild and Scenic Rivers," *Technical Report of the Interagency Wild and Scenic Rivers Coordinating Council* (Washington, D.C., National Park Service, 1998).

Andy Kerr, *Oregon Wild* (Portland, Oregon Natural Resources Council, 2004).

Samuel C. Lancaster, *The Columbia—America's Great Highway Through the Cascade Mountains to the Sea* (Portland, J.K. Gill Co., 1915).

CHAPTER 4
WILDERNESS ISSUES IN THE NATIONAL FORESTS

Three Sisters Wilderness

Of all of Oregon's wilderness areas, the Three Sisters is the most storied. Its saga encapsulates the main lines of the stories about all the rest—dealing with threats to wild country from logging and logging roads.

In the 1920s, Willard Van Name of the New York Natural History Museum, in a book he wrote, proposed that this scenic area in the central Oregon Cascades be made a national park. Across the West, the Forest Service faced similar proposals that its holdings be turned over to the fledgling National Park Service for new national parks. It did not want to lose land on a wholesale basis.

As a counter move, the Forest Service began setting aside swaths of its land as Primitive Areas that were far from improved roads. Most of these had little timber of value. Their boundaries were so loosely drawn that some of them even had low-class roads in them.

In this manner, a 191,000-acre Three Sisters Primitive Area was set aside in 1937, even though some residents of Eugene and Bend preferred a national park.

To further buttress the appeal of this alternative, the next year the Forest Service's Chief of Recreation, Robert Marshall, who had just visited the area, added another 50,000 acres to the Primitive Area. This tract of land was on

Three Sisters from Lowder Mountain

the west side of the Primitive Area and was heavily timbered; it was west of Horse Creek. Marshall later became recognized as a great champion of wilderness, but died of heart failure in 1939 at only thirty-eight years of age.

Around 1937, Marshall decided that the Forest Service should study all of its Primitive Areas to select better boundaries for them and to exclude all roads and inappropriate development. Under regulations U-1 and U-2, it would reclassify its Primitive Areas as either Wilderness or Wild Areas ("Wild" was a designation for smaller areas). Most of its studies to reclassify Primitive Areas, however, did not get under way until after World War II.

When the Forest Service turned its attention in the late 1940s to the reclassification of the Three Sisters Primitive Area, the timber industry was intent on gaining access to most of the timber in the national forests. It pressed the Forest Service to minimize the amount of merchantable timber put in wilderness areas.

The Forest Service obliged them by proposing to eliminate the forested area west of Horse Creek that Bob Marshall had put in, arguing that this area was not especially scenic. In compensation, the Service proposed designating new wilderness in the high country around Mt. Washington and Diamond Peak, which had little merchantable timber.

In 1950 and 1951, the Service took various interested conservation groups and academics on tours to examine the areas involved. But those that went were not all of one mind. Some agreed to deletion of the area west of Horse Creek (though wanting it to be studied more), while others completely opposed the deletion. Members of the Eugene Natural History Society and the academics were most opposed.

Thereupon, they formed a new group to give voice to their opposition: the Friends of the Three Sisters Wilderness. The dean of men at the University of Oregon, Karl Onthank, and his wife Ruth, played major roles in this organization's defense of the 53,000 acres, which everyone thereafter called the timbered area west of Horse Creek. Onthank had earned a reputation as a conservation leader with widespread interests. Biologist Ruth Hopson also was quite involved.

In 1954, the Forest Service issued its formal proposal embodying its proposal of the preceding years. At public hearings on this proposal, about two-thirds of those testifying opposed the deletion, while about one-third supported it. Those supporting the proposal comprised mainly people connected to the timber industry.

In the years since the initial inspection tours, the Federation of Western Outdoor Clubs had changed its position, deciding now to oppose the deletion. It was a group that included the Obsidians of Eugene, the Chemeketans of Salem, and the Mazamas of Portland. Earlier, it had equivocated. These opponents were supported by Senators Richard Neuberger and Wayne Morse, as well as recent Forest Service Chief Lyle Watts.

As one might expect, the timber industry supported the Forest Service position. Also supporting the Service was William B. Greeley, who had been chief in the 1920s and later headed the West Coast Lumbermen's Association.

The clash drew the attention of national conservation groups. It was the first showdown over reclassifying a Primitive Area. Some of their leaders, who testified at the hearings, now concluded that it was risky to rely on the discretion of the administering agency to protect wilderness. This sobering experience convinced them that a new law was needed to protect wilderness—a Wilderness Act. A bill to make that Act a reality was soon supported by both of Oregon's senators, Neuberger and Morse, who were among its initial sponsors.

In February of 1957, following the presidential election, the Forest Service announced its final decision. Hardly any changes were made following the hearings, except that there was to be more study, prior to logging, of small units in the 53,000 acres that they called Special Areas. I helped Karl Onthank craft petitions for these areas, and eventually five of them were set aside there: a botanical area, two geologic areas, and two scenic areas.

The frustrations over the loss of the 53,000 acres revived interest in the idea of a national park in the central Oregon Cascades. In October 1959, David R. Simons of Springfield wrote an article for the *Sierra Club Bulletin* proposing the specifics of such a park. A new statewide group was formed to promote it: the Oregon Cascades Conservation Council. For a while in the early 1960s, I staffed it. Then in the late 1960s, the national Sierra Club injected some new energy into efforts to promote this idea, with Larry Williams of Portland doing the promotion. But the Club put its principal efforts into promoting a new national park in Washington state in the North Cascades, where it had a more energetic partner, the North Cascades Conservation Council. In 1968 that park was established.

These failures eventually fueled a revived campaign by conservationists on behalf of the surviving, southern portion of the deleted area—that around French Pete Creek and to its south. That campaign drew greater support and produced a different result. Its story will be told subsequently under that heading.

In concluding, I should note that in the early 1960s, I undertook a different challenge involving the Three Sisters Wilderness. The U.S. Pumice Company had filed location notices in 1963 to open a quarry to mine block pumice in the wilderness at Rock Mesa near the South Sister. The development entailed building an access road into the Wickiup Plains. Facing this threat, I devised a strategy on behalf of environmentalists to fend it off.

In looking through the Code of Federal Regulations, I noticed provisions for appeals that were available to all aggrieved parties who were adversely affected. Since we were aggrieved, I thought, why could we not use this provision? I imagined this provision had heretofore been used just by commercial interests, but I saw no reason that others could not use it as well.

So in 1963 I filed a formal appeal to contest these claims, arguing that their procedures were defective, for various technical reasons, in filing for placer claims. I doubted that they were qualified to have 160-acre association placer claims rather than ordinary twenty-acre claims. I urged the Forest Service to investigate the circumstances. I was probably the first to use this appeal provision for conservation purposes.

In due course, this appeal was rejected on the grounds of insufficient evidence, though I had asked that the Forest Service and BLM[17] investigate and gather evidence. I later learned that, behind the scenes, both the Forest Service and the BLM had opposed the project and that the BLM agreed that the 160-acre claims were not justified and that there were other irregularities (e.g., a third had no minerals; the legal descriptions were faulty, etc.). For another two decades, a contest over the claims ensued both in the open and behind the scenes.

In 1977, the Forest Service itself appealed this ruling to the Interior Board of Land Appeals, which ruled adversely in 1981. While the Forest Service decided not to contest this matter further, the state Sierra Club, ONRC, and Friends of the Three Sisters Wilderness did. The Club found a lawyer to pursue this contest, with the necessary funding, and launched lobbying of its legislators. But in 1982, Congress finally decided to bring the controversy to an end and resolve the issue by providing funds to buy out the claims they felt were valid (about half of them); they accomplished this the following year. While the entire delegation backed this move, the lead was taken by the members on the appropriations committees: Senator Hatfield and Representative Les AuCoin.

The long appeal process had bought time for opposing public opinion to become sufficiently aroused to prompt Congress to act.

So not all the news from this time was disappointing.

---

17 The BLM is in charge of administering the mining laws on the public domain, including the national forests.

First Reform in Federal Mining Laws

Another mining issue affecting wild lands arose in Oregon a decade before the Rock Mesa issue arose.

For years, the General Mining Law of 1872 had been interpreted to allow mining claimants to cut timber on their claims and build summer homes. It had also been used to hamstring the Forest Service in managing surface resources. A claim in the Rogue River National Forest in the mid-1950s—the Al Serena claim—mushroomed into a political issue. A patent had been issued by Interior Secretary Douglas McKay on a claim in which Senator Richard Neuberger charged than the claimant "had never mined so much as a thimbleful of ore" from it, while living on it. In due course, that led to congressional hearings on the matter.

The upshot was enactment of the Multiple Use Mining Act of 1955, which gave the Forest Service control over the surface resources of unpatented claims so long as legitimate mining activity was not impeded. It put greater emphasis on proving the validity of claims and helped clear away stale claims that had accumulated. The Act also removed common varieties of minerals, such as sand and gravel (but not block pumice) from being claimed under the 1872 Act and put their disposal under the Materials Act of 1947.

This colorful episode in Oregon politics demonstrated the potency of conservation issues here. At the national level, it constituted the first real reform in the Mining Law of 1872. It is worth remembering.

Minam River Valley

One of Oregon's most spacious and spectacular wilderness areas is the Eagle Cap Wilderness in the state's northeastern corner. With mountainous scenery reminiscent of the Rocky Mountains, it anchors the surrounding terrain. It was made a Wilderness Area in 1940 by the Forest Service (previously having been a Primitive Area) and became a part of the statutory wilderness system in 1964. At 358,000 acres, it is Oregon's largest.

It was not so large, however, when it was added to the statutory system, and it had few forested valleys at lower levels. But outside its boundaries along its western flank, one did exist—the fifty-mile long Minam River valley. In 1960, it was still free of roads, with only a few dude ranches in it. But its future was then in doubt. The Forest Service was being urged to build an access road up the valley and let bids for timber sales.

U.S. Supreme Court Justice William O. Douglas knew the area, having been married for a while to a woman from La Grande. Friends there were keeping him informed, and he was alarmed at this prospect. Knowing David Brower of the Sierra Club in San Francisco, he urged him to rally public opinion to save the valley from this fate.

Brower responded by laying plans to put out a flyer appealing for public support to keep the area roadless. He arranged for photographer Phil Hyde to go there in the summer of 1960 and asked me to accompany him and write its text. We stopped on the way to visit the office of the Wallowa-Whitman National Forest to gather information. Then we rode horses into Red's Horse Ranch to stay. The following day I walked eighteen miles to gather my impressions. Hyde drifted along behind, slowly capturing interesting shots. Soon thereafter we gave Brower our material, and he put out a flyer mailed to West Coast activists, with an appeal to "Stop This Crazy Cat."

This began a twelve-year debate over the future of this valley. No immediate decisions were made to put Caterpillar tractors on the ground for road building. I testified a few months later at an agency hearing in La Grande. Dr. Charles Quaintance, a biologist at Eastern Oregon College, rallied opposition at that time. In 1962, I secured a stay on moving ahead with development through a process I will describe in the next section. And later the ranks of the opponents grew, as also did the intransigence of the local timber firms who wanted the road.

But the wilderness supporters had a formidable friend. In this case, it was Senator Mark Hatfield, who in 1971 introduced a bill to give wilderness status to the valley. He managed to arrange a field hearing in La Grande in the summer of that year, where people turned out in numbers that extended as "far as the eye can see." On the 4th of July wilderness backers even marched in Joseph to show their support, carrying coffins lamenting what the Minam would have without wilderness.

But the contending forces were sharply divided over adding the valley to the adjoining wilderness. Those who opposed wilderness and wanted to log the area also had powerful support—the local congressman, Al Ullman (D.), and the governor, Tom McCall (R.). However, in this case Senator Hatfield proved to be the better tactician. He worked closely with Senator Henry Jackson of Washington state to steer around Ullman and his House supporters, using shrewd strategy.

At last, in 1972 he succeeded—the 72,400 acres in the valley (downstream to the national forest line) were added to the Eagle Cap Wilderness.

In passing, I should note that Hatfield was unpredictable on questions of wilderness and timber, his stance varying with his sense of how much, at a given time, he needed to cater to environmentalists. While McCall is now thought of as environmental champion, his green streak did not extend to timber issues. And Ullman never embraced environmentalism, though he thought of himself as a liberal.

Waldo Lake Basin: Thwarting Logging

At the same time I was working on the Minam, I had also joined Karl Onthank, a longtime leader of local conservationists, of Eugene in trying to protect Waldo Lake. Waldo Lake had just lost its status as a Limited Area (a temporary protective designation of the northwest region).

North of the Willamette pass, Waldo Lake is Oregon's largest glacial, mountain lake and the lake with some of the world's clearest water. It lies in a shallow basin at upper elevations with little disturbance—until I discovered plans to log forests in parts of it.

At first we were intent on combating plans to build a new, paved road into the basin, but then I discovered plans to eventually let timber sales in it. In examining multiple use planning maps of the Forest Service, I noticed that not all of the high country was earmarked for recreation by being shaded

Waldo Lake, at South End

in orange. There were places where the green shading extended into the high country, which indicated places where logging could occur under the rubric of multiple use. This was the case up and down the Cascades, but one of them was at Waldo Lake.

We were assured that no immediate plans existed to log there, but I soon learned that this would inevitably occur once their calculations of the allowable cut counted on these forests as part of the timber base (later known as the "allowable cut effect"). Without this timber, they would be cutting timber at too fast a rate.

Getting nowhere with efforts to persuade the Forest Service, I decided to appeal for help to prominent northwest senators. Because I had good connections with them, I appealed to Senator Wayne Morse and Senator Henry Jackson; they were then the senior senators from Oregon and Washington. I asked for a hold on development at Waldo Lake and in the Minam. Now that I was working for conservation groups, I also asked for a holds on development in the North Cascades and at Cougar Lakes, a bit to the south of it. Justice Douglas was concerned with both of these too. I hoped that Forest Service plans for all these places could be reassessed.

Both senators agreed and appealed to the Secretary of Agriculture to have the Forest Service delay their plans and re-evaluate them. In the fall of 1962, the Secretary's office responded by issuing a broad new policy direction for their high country in these states. The Forest Service called it their High Mountain policy. In it, they decided to try to quiet these controversies by withdrawing plans to log in the high country (usually above 4500 feet). That zone would be devoted entirely to recreational use—now to be entirely shaded in orange on their maps, with the fingers of green removed.

Who would have thought that this local controversy at Waldo Lake could morph into a major policy change? It did not end all of these controversies, but now at least the high country would be spared the ravages of major logging.[18]

Later generations of environmentalists thought this was a deliberate effort to "buy us off" by allocating the high country to us, while allocating the lower, heavily timbered country to lumber companies. Alas, too many conservationists did not understand that once even the high country was not secure. We had to fight for it.

Much later, in 1984, 39,000 acres on Waldo Lake's north and west sides (where logging was originally slated) were designated as wilderness.[19] That story will be told as part of the French Pete Creek story. By keeping logging out of the lake's basin, we were able to protect its exceptional clarity. Today additional protection has been accorded to the lake in other ways: the state has made it a state Scenic Waterway that aims at protecting its natural setting and water quality. And the state has banned seaplanes and boats using combustion motors.

---

18 It is not clear how long this policy persisted, particularly around such places as Mt. Hood.

19 A trail around the lake has been named for Representative Jim Weaver, who championed wilderness here and elsewhere.

Mt. Jefferson Wilderness

The Mt. Jefferson Wilderness in the central Oregon Cascades is one of the most scenic and accessible wilderness areas. And at 10,495 feet, Mt. Jefferson is a real mountain with glaciers and faces that require rock climbing; I have done it a number of times. But the wilderness area lacks breadth; it is only seven miles wide in most places. At its northeast edge, the land accessible to the public is even more limited. The Warm Springs Indian Reservation boundary runs generally from the east to the crest just south of Mt. Jefferson. At the time (prior to 1970), this fact made the hiking public even more conscious of the overall narrowness of the Primitive Area.

Environmentalists made repeated efforts to overcome these shortcomings.[20] But they were frustrated by the way the Forest Service had engaged in what I then characterized as "leap frog" logging. In the late 1950s and early 1960s, the Service ran access roads (which were primarily for logging) right up to the boundaries of the Primitive Area and then let timber sales near the ends of the roads. Miles of uncut forests were bypassed. Over time, then they would let sales elsewhere along the roads.

In this way, they fixed the boundaries on the west and east sides before reclassification even began or public comments were invited. Options were foregone before studies were even commenced. The most egregious example was on the west side along Whitewater Creek.

In 1962, I even proposed to Congress that they should require that public hearings be held before the Forest Service let timber sales within two miles of the boundary of a primitive area—to no avail.

The agency planned to reclassify the Mt. Jefferson Primitive Area in 1963 before the Wilderness Act was even passed. So in 1963 I studied the area and, in a thirty-eight-page report, and on behalf of the Federation of Western Outdoor Clubs, proposed boundaries for the new wilderness of 117,000 acres. At that time, we did not believe there was any chance that logged and roaded areas would be accepted—even if they were restored. It is worth

---

20 They never challenged the status of the Indian Reservation nor wanted to.

Mt. Jefferson from Jefferson Park

remembering too that, at this time, this was still an agency process where the Forest Service made all the decisions. So I did not include any of them.

But I did find another mile of undisturbed forest land that could be included on the west side along Pamelia Creek. It afforded a very welcoming environment for hiking. That proposal was accepted.

As it turned out, Mt. Jefferson was not among the units initially included in the new statutory wilderness system. Further processes would need to be undertaken to include it.

In the next few decades, environmentalists changed their strategy by attempting to have damaged areas included and restored. They were outraged that the latitude of Congress's decisions had been preempted by unilateral logging commitments. They did not want to reward "bad" behavior on the part of the Forest Service.

Not all environmentalists agreed with that approach because it undermined efforts to keep undisturbed areas from being logged so that they could qualify as wilderness. If it was such a simple matter to restore them, keeping them intact would not be a matter of great moment. Thus, this approach was something of a double-edged sword.

CHAPTER 4—Wilderness Issues in the National Forests

Former Governor Charles Sprague of Salem, who was a hiker who knew that area well, did not have any enthusiasm for the new approach. He was a member of an outdoor club that had supported the Wilderness bill.

He was joined then by Senators Morse and Hatfield, who also did not favor adding another 30,000 acres with as much as 11 million board feet of merchantable timber, supposedly costing 600 jobs. The new proposal of the environmentalists in 1968 called for an area of 125,000 acres. This proposal, however, did draw heavy support at hearings in 1966 in Salem. But that was not to be.

Gradually over time, though, the size of the preserved area crept up. Having been first set aside in 1930, it was expanded administratively in 1933 to 83,000 acres, and then in 1963 to nearly 97,000 acres, then in 1968 to 100,000 acres, and finally in 1984 to 107,000 acres. In that year, Oregon Wild was instrumental in having 6800 acres of forested areas added. The Sierra Club, the FWOC, and the Oregon Environmental Council (OEC) were active in securing its initial level of protection in 1968.

The Forest Service was also piqued over the flexibility of the environmentalists on questions of non-conforming intrusions in wilderness. Included in the 1968 boundary adjustment was the 3000-acre Marion Lake enclave. That area then was used for boating and even had a boat house there. The Forest Service advised against its inclusion but Salem conservationists wanted it included.

The Forest Service thereupon ordered that, within one year, the boats and boat house be removed. This abrupt order upset Senator Hatfield, who had language put in a committee report emphasizing that non-conforming uses could be removed over time in a gradual and phased manner.

It was amazing how much conflict and ill-humor this beautiful area could inspire. And this was not the end of it. More was to come at nearby Opal Creek, which will be covered later.

Wilderness Act Inclusions

The Wilderness Act, which was enacted in 1964, basically established a national system for the protection of areas that Congress would designate over time as wilderness. But nine million acres of reclassified areas were put in the system at the outset, and these included some areas in Oregon. In fact nine Oregon units were put in the system then, totaling 662,000 acres.

The largest were the Eagle Cap Wilderness, the Three Sisters Wilderness, and the Kalmiopsis (south of the lower Rogue). The others were Strawberry Mountain, Mt. Washington, Diamond Peak, Mt. Hood, Mountain Lakes, and Gearhart Mountain. While the reclassification of the Mt. Jefferson Primitive Area had progressed, it had not been completed in time.

A number of these areas have been expanded over time, particularly the larger ones.

A note in passing about Mt. Hood. One might wonder why its wilderness is not larger since it is such a significant place, being Oregon's highest peak and having been proposed as a national park. But wilderness opportunities there are limited by the nearby ski areas with their lifts, as well as by lodges and roads, particularly on the south side of the mountain.

However, areas have been added to it over time, though often in an odd, disjointed manner. Now less emphasis is placed on compact boundaries and more on saving remnants of wild land (see Chapter 9).

Conservationists have long discussed the question of whether Oregon has less wilderness than it should have, comparing it to other West Coast states. It does, but a number of factors are worth remembering.

This is a state that was long dominated by a powerful timber industry that stood against much wilderness, particularly forested wilderness.

Moreover, the nature of the topography in Oregon often constrains the size and shape of the wilderness areas here. In designing wilderness areas, environmentalists usually try to enclose mountain masses that are roadless. In other nearby states, these mountain masses often have breadth in alp-like terrain, as in the North Cascades and the Sierra.

In the Oregon Cascades, however, the mountain masses are likely to be long and narrow. While environmentalists usually try to broaden them into the abutting forests of the "old Cascades," the lateral forests are contested terrain, and often have already been logged and roaded. The shape of the wilderness areas reflects both these contests and the nature of the terrain.

CHAPTER 4—Wilderness Issues in the National Forests

But where we have alp-like terrain, as in the Wallowas, we tend to have breadth as well as length, as with the Eagle Cap Wilderness.

And finally, one might suggest that Oregon has a greater variety of formulas for preserving natural values in situ. It has been very inventive in finding ways to preserve natural values. It has much to be proud of.

This book will continue to set forth the case for our variety and inventiveness in preserving nature.

References

Andy Kerr, *Oregon Wild; Endangered Forest Wilderness* (Portland, Oregon Natural Resources Council, 2004).

Kevin R. Marsh, *Drawing Lines in the Forest: Creating Wilderness Areas in the Pacific Northwest* (Seattle, University of Washington Press, 2007).

Michael McCloskey, *In the Thick of It* (Washington, D.C., Island Press, 2005).

Lawrence C. Merriam, *Saving Wilderness in the Oregon Cascades: The Story of the Friends of the Three Sisters* (Eugene, Friends of the Three Sisters, 1999).

Willard Van Name, *Vanishing Forest Reserves: Problems of the National Forests and National Parks* (The Gorham Press, Boston, 1929), part III, chapter V.

CHAPTER 5

OREGON'S ENVIRONMENTAL LAWS: KEY PROGRAMS

Oregon's Beaches Protected at Last

Once it became clear that only the wet beaches[21] were public beaches, public attention focused on how they would get to those beaches. In the 1960s, various solutions were put forth: buying the adjoining dry sands, zoning controls, or trying to vest ownership in the public up to various lines; e.g., the vegetation line, or to a given elevation level, such as sixteen feet above mean high tide level.

But a real storm of controversy broke when developers began to buy up coastal frontage and deny access. William Hay did this in both Pacific City and in Cannon Beach, where he owned hostelries.

In 1967, Governor McCall put forth a hopeful solution drafted by the Parks Department. The legislature then debated the beach bill, with heavy media coverage, fierce opposition, and ten House hearings. Finally, the legislature fashioned a compromise that tried to fix a clear line where the public's easement ended and private property began. It opted in favor of the sixteen-foot elevation level, which was where the vegetation line was often found. This approach used the zoning concept.

The legislature declined to embrace the better option of granting the public rights to cross the dry sands, using common law property concepts.

---

21 Those swept by tidal fluxes.

Site of Beach Encroachment Controversy at Cannon Beach

But Hay thought even this weaker approach involved an unconstitutional taking, challenging it unsuccessfully in the state courts.

His challenge, however, convinced many conservationists that this enactment was vulnerable and that it would be better to amend the state's constitution to guarantee public access.

But how to frame the constitutional remedy?

Dr. Robert Bacon thought it would suffice to merely guarantee free access by the public to the beaches, though these public rights would only be established if adjoining property owners did not file suit within a year to contest this access. He did not deal with the issue of already established rights of public access.

State Treasurer Robert Straub and Janet McLennan headed up a competing group that took a different approach, calling their group Beaches Forever. Their approach confirmed rights of public access to wet sands and said that it ran from the vegetation line on out across the dry sands. In the limited number of cases where beach ownership had already moved into private hands, their measure would have authorized the state to acquire them, as well as routes of access to reach them. For four years, a one cent per gallon tax would have been imposed on gasoline to pay off bonds to buy this land; $30 million could be spent on beach acquisition.

The Straub and McLennan group had more experience in running campaigns and got off to faster start. They quickly got their ballot title (on May 1, 1968) and gathered the necessary number of signatures to get on the ballot. They enlisted teams of volunteers to gather these signatures, raised funds, and gathered support from conservationists and outdoorsmen. Their measure qualified as Proposition 6.

Bacon could not compete. He was slow to get his ballot title and gather signatures, notwithstanding support from Gov. McCall, who had dramatically confronted Hay on May 13, 1967, when he tried to block access. McCall eventually supported Straub's measure when it was clear that it was the only one that would qualify. Senators Packwood and Morse already supported it.

While Straub ran a campaign with widespread involvement, including by school children, he encountered formidable opponents. It came from the heavily financed Oregon Highway Users Conference, which Straub asserted was a front for the oil companies. Political operator Ken Rinke had persuaded six oil companies to finance a massive, last-moment media campagin to defeat Straub's measure. That group stressed that the measure would have imposed a new gas tax, and asserted that it was all unnecessary because the public already owned the beaches. They used the tag line: "Beware of Tricks in Prop. 6."

On Election Day in 1968, their measure went down to defeat. The oil company campaign was very effective. Ironically, following their victory, the companies soon raised their prices by a cent a gallon. And Straub pointed out that post-election reports showed that no Oregonians were listed as contributors to Rinke's campaign.

But the public's rights were soon vindicated in another fashion. In December of 1969, the Supreme Court of Oregon at last ruled on the issue implicit in Hay's challenge to the action taken by the legislature in 1967—now in the form of a suit by the state attorney general against Hay to force him to remove his fence on the dry sands. Acting unanimously, in an opinion written by Alfred T. Goodwin, it ruled that the state had a right to protect the public's recreational easement to reach the wet sands by crossing the intervening dry sands. It proclaimed that the state even had the power to prevent adjoining owners of private property from blocking access to their dry sands so to deny public access to the wet sands.

It said the public had gained a right of continued access under the ancient English common "law of custom." This doctrine could apply to broad areas, not just specific places—in this case to the beaches from the northern to the southern border of the state. In this case (*State ex rel. Thornton v. Hay.* 462 P. 2d 671-2, OR 1969), the court insisted that "from the time of the earliest settlement to the present day, the general public has assumed that the dry-sand area was part of the public beach." Crossing these dry sands to reach the wet sands had been "an unbroken custom" by the public from the time of earliest settlement. All of the requirements set forth by Blackstone to apply this doctrine had been met in this case. It noted also that state officials had policed these beaches and collected litter on them, indicating that the state had long acted, without challenge, as if it had rights to do this.

This was a landmark case in state conservation law. It was later interpreted to mean that it did not apply to all beaches, such as those beaches in isolated coves that the public could not easily reach. And when the ruling was again challenged in later years (1993), the court held that exclusive use of the dry sands was not part of the "bundle of rights" that the owner acquired when purchasing the land.

This expanded easement was now vested in the state, as a trustee for the public. The state soon designated all of this as a new state recreation area. Surely all this public interest campaigning had emboldened the Supreme Court's justices. Os West's vision was at last fully realized. But this vision will need continued vigilance.

Bottle Bill

Growing interest in environmental subjects at this time prompted a number of Oregonians to start paying attention to litter in public places. One of them was Richard Chambers of Tillamook, who had been picking up roadside litter, and noticed that cans and bottles were everywhere. Portland attorney Al Hampson also became concerned about all the litter on Oregon beaches, which he noticed especially around his beach house at Neskowin. And Portland businessman Don Waggoner also became concerned about the growing problem. Having at first headed a group involved in recycling campaign signs, he began to pay attention to the components of prevalent litter.

At the end of the 1960s, Waggoner headed a group in the Oregon Environmental Council (OEC) that called itself "People Against Non-Returnables." It now included Richard Chambers, though he also continued to work by himself.

Waggoner felt the need to gather more facts; his group visited bottlers, brewers, and grocers. They then did a survey of roadside litter in the leading cities: Portland, Salem, and Eugene. Among other things, they found that 157 million containers became solid waste in 1969 in Oregon, and that cans and bottles constituted forty percent of the litter along roadsides. All of these findings went into a report they issued for OEC, which included graphic photographs.

An abortive effort, which would have imposed an outright ban on beer and soda containers that were non-returnable, was first made in the legislature in 1969. It grew out of an attempt by Richard Chambers, who got his state representative, Paul Hanneman, to sponsor the bill. Chambers did make an effort to mobilize support, with letters coming in to legislators and witnesses testifying for it. But it fell short, facing fierce opposition from bottlers, fabricators, and grocers. Unions opposed it as well, fearing a loss of jobs.

A sufficient effort had not been made to organize widespread support. Also Hanneman could not persuade Republican Governor McCall to line up support among fellow-Republican legislators. Moreover, opposition did not abate when Hanneman offered to substitute a deposit instead of a ban.

Opposition from beverage distributors was particularly intense. And the grocers worried about what it might mean for them. They were concerned about being overwhelmed by returned containers, including ones they had not sold. They feared they would not have enough space to store them. And they did not want to undertake what they viewed as a "messy" job.

But the legislative battle in 1969 did lead to a study by an interim committee, which held sixteen hearings and heard from as many as 143 witnesses. As the upshot, a package of nine bills was delivered to the next session. One of them was the five-cent deposit bill finally enacted in 1971, which included a provision requiring that these beverage containers be returnable. It also included a ban on detachable pull-tabs, which had come into use in the 1960s and become a problem. This shift in how tabs work has now become worldwide.

In 1971, the lineup of opponents had not changed, but one soft-drink bottler did support the measure. The Dr. Pepper dealer thought they would save money if bottles were returned for reuse, since the container cost more than the product. But most of the bottlers felt it would depress sales.

But this time, the supporters were much better organized. They made heavy mailings to groups that might support the measure, with crews of volunteers doing the work. Aroused cadres of citizens buttonholed legislators in Salem, stalking the halls continuously. More groups now joined in supporting it, including the Granges. Many hearings were held in both the state House and Senate.

OEC tried to work out reasonable compromises with affected interests. For example, it worked out an arrangement with Bill Wessinger at the Blitz-Weinhard brewery of Portland to get his support. They reduced the amount of the deposit on smaller beer bottles, the so-called "stubbies."

Portland attorney Al Hampson arranged these meetings with Wessinger, who was a friend of his. He also raised the funds needed to hire a professional lobbyist, who had the experience to get the bills assigned to the best committees. He also got his friend, Lee Johnson, who then was the state's Attorney General, involved in putting forth ideas about how the bills might best be framed. Johnson, who then formed his own group ("Citizens Against Litter"), later testified in his private capacity and tried to influence key legislators.

CHAPTER 5—Oregon's Environmental Laws: Key Programs

Lobbying by both sides was intense, with recognition that this could be landmark legislation. It even gave rise to charges of attempted bribery. At that time, State Senator Betty Roberts chaired the Consumer Affairs Committee, to which it had been referred after being passed out of another committee. Her committee joined in reporting it out favorably, but on the floor of the Senate, she charged that an industry lobbyist had dangled the prospect of substantial financial support for her next campaign if she would vote against it. She did not.

The climate was so poisonous that Governor McCall even denounced the container industry for its campaign of what he characterized as the "four D's: Distortion, Deceit, and Dollars, equaling Defeat" ("equaling" meaning the defeat that opponents hoped to engineer). National media later quoted Attorney General Johnson in decrying the character of the lobbying. Even the Association of Oregon Industries (AOI) complained about it.

Ultimately, however, the measure passed in both houses. While the margin of passage in the House was ample, it passed the Senate by only 16–14. If Senator Betty Roberts had not held firm, it would not have made it. After passage, it faced challenges to its constitutionality in the courts, but the courts upheld it.

The program has been quite effective in achieving its purpose, with cans and bottles now comprising only four percent of the roadside litter (in contrast to forty percent before). The overwhelming majority of bottles are now returned. Moreover, the vendors of beverages benefit by being able to keep the deposits on the seven to fifteen percent of the deposits that are never claimed. This is a tidy windfall. In many other states, unclaimed deposits revert to the respective states.

In successive years, containers for other beverages were added—for water and flavored water in 2007 and for juice, coffee, tea and, energy drinks in 2011. These efforts were spearheaded in the legislature by Representative Vicki Berger, who is the daughter of Richard Chambers. In 2011, raising the deposit fee from five to ten cents was also authorized if the redemption rates fall below eighty percent for two years in a row, which has now happened. Also experiments are to be undertaken with redemption centers that beverage distributors will organize in the form of a cooperative collection company. Three have now been built, and five more have been planned; so far the experiments have been encouraging.

Not all efforts to expand and modernize the bottle bill have succeeded. A broadly framed ballot measure was put forth in 1996 by OSPIRG, but went down to defeat in the face of heavy opposition from the beverage industry and many grocers (not including Fred Meyer). Little money was raised to promote this measure.

Over the years, eight others states enacted similar deposit laws, with limited variations in two others, and in several cities and counties.

Oregon continues to enjoy cleaner public places because of this landmark law.

In later years, many felt that plastic shopping bags also contributed to roadside litter. Finally, in the period 2010–2012, cities began to prohibit their use by stores. Portland, Eugene, and Corvallis did this. Earlier efforts had been made to get them recycled and for them to be made of decomposable materials; these enjoyed modest success.

Oregon's Land Use Law

In the late 1960s, many environmentalists felt something had to be done to stop sprawl and unplanned development. A half million acres in the Willamette Valley had been lost to development since the mid-1950s. Books and speeches by planner Ian McHarg had fueled notions that comprehensive land use planning was needed to deal with the problem—particularly planning embodying environmental constraints. Most states only had such planning and zoning in cities but not in rural areas, which were under the control of counties. While Oregon had permitted its counties to do such planning since 1947, few did. Did Oregon want to mandate such planning?

Governor McCall thought that Oregon should, and he pressed for this approach. He felt the situation here was dire. He denounced "despoiling… [the] land" by "grasping wastrels" who were propagating "sagebrush subdivisions, coastal condominia," and spoke of the "ravenous rampages" of development in suburbia.

He said local government was not standing up to these pressures, with its response being "timid, fragmented, disjointed, reactive, and pitifully inadequate."

Instead, he proclaimed that we needed a statewide system of land use planning. At the start, his proposal was embodied in SB 10, which OEC supported with testimony. It was enacted in 1969 with little resistance. The measure required counties, as well as cities, to prepare comprehensive plans and zoning codes. But few took the requirement seriously and developed plans. Under this limited law, the state did not provide any funding for the counties to develop such plans, nor did it provide any effective mechanisms to enforce the requirement.

The very idea of this kind of program was offensive to many actors—builders, real estate interests, and rural property owners. They soon put an initiative measure on the ballot to try to repeal it, but urban interests united to defend it. Both labor (AFL-CIO) and industry (AOI) joined in support of it, the latter urged to do so by both John Gray of OMARK Industries and Glenn Jackson of Pacific Power and Light. They were joined by the League of Women Voters, the Portland City Club, and OEC, which made major mailings to rally voters. The repeal effort went down to defeat by a substantial margin, with only forty-four percent favoring repeal and fifty-six percent voting to retain it.

The land use program's backers then realized that they needed to do more to build the case for the program and to broaden popular support. At the behest of Secretary of State Clay Myers, planners in the Willamette Valley took the lead in organizing "Project Foresight." Under the guidance of Lawrence Halprin and associates, it put forth various possible scenarios for the future in an environment without planning.

At meetings around the valley, they gathered input on people's concerns about how unchecked development would overrun the valley's farmland without planning and zoning. Over 270 meetings were held, with as many as 20,000 participating. Many were outraged over the impact of the huge

Charbonneau development recently constructed on prime farmland (class 1) near Wilsonville. HUD provided some of the funding for these sessions.

All of this was pulled together at a subsequent conference in Portland, with 600 participating, including many businessmen. OEC arranged for Ian McHarg to address them.

McCall also rallied public support, campaigning all over the state. He also had his Health agency do a study of the health implications of runaway subdivisions. In looking at recent ones on 85,000 acres across the state, it found that one-quarter of them had inadequate arrangement for handling sewage. In face of this finding, McCall said this neglect was inviting a "serious public health problem." "Irreversible environmental damage is being done," he proclaimed.

McCall found an effective ally in Republican State Senator Hector Macpherson, a Linn County farmer. Some farmers in his district were beginning to feel hard pressed by spreading housing tracts. These concerns prompted agriculture extension agents there to seek research on soil types of special value as farmland. The state had already experimented with tax breaks for land devoted exclusively to agriculture.

In between the 1971 and 1973 sessions of the legislature, they began work on the development of a more comprehensive approach to land use planning, with a stronger role for the state. Democratic State Senator Ted Hallock of Portland soon joined him in sponsoring the new bill and helped guide it through the legislative process. This approach, embodied in SB 100, established a state Land Conservation and Development Commission, a counterpart agency, and a staff to administer the program.

At hearings, OEC and the Sierra Club supported the new bill, as did the League of Women Voters, but all asked for amendments to make it even stronger. OEC wanted it to be better aligned with pending federal land use legislation (never enacted), and the League suggested better provisions for public participation.

But the detractors were as vocal as ever, with the realtors asserting that no need had been shown for change that undermined local control. Others asserted that police powers would be used for ends that were essentially esthetic. Timber interests either advised "going slowly" or were outright opposed. Others spoke of the dangers to private rights and denounced "socialistic planning." While many farmers supported it to preserve farmland and the agriculture industry, others organized something called the "Farmers Political Action Committee" to oppose it.

This opposition managed to bring movement in the legislature to a halt. SB 100 could not even get enough votes to be reported out of committee.

At this point, Macpherson and Hallock knew the moment of truth had arrived and that something different had to be done. What Hallock did was to ask former legislator L. B. Day to head up a task force to negotiate enough compromises to put the bill in a form that could get passed. Day was a teamster union official who had headed the Department of Environmental Quality and was widely admired (as well as feared) for his political acumen. McCall encouraged Day to undertake this job. The League of Oregon Cities helped him forge a compromise.

Day did manage to complete his task, with the Senate approving his version. However, OEC and OSPIRG and other environmentalists felt that Day's bill had too many compromises. Fearing that these misgivings would cause the House to strengthen the bill and thereby jeopardize its re-approval when it went back to the Senate for concurrence, Hallock convinced the House Committee (headed by Representative Nancie Fadeley) not to amend the bill. They heeded his advice so that Day's version was finally approved by the House and signed by McCall in May of 1973.[22]

Eventually OEC decided to support SB 100 and urged McCall to sign it. They hoped that future legislatures would cure its shortcomings. The environmental organization asked McCall to pledge to work for their enactment, but he refused. Nonetheless, OEC urged it be signed, with State Senator Macpherson thanking them for finally coming around. Macpherson is widely acknowledged to be the father of the state's land use program.

The 1973 session of the legislature also enacted a number of other land use–related laws, including ones governing city and county planning, subdi-

---

22 Curiously, at one point McCall actually claimed that Day's version was even better and cured all the defects in the program that SB 10 had established.

visions, agricultural uses, and open spaces. These strengthened implementation of SB 100.

The enactment of SB 100 and related measures put Oregon in the vanguard of environmental leaders, though there are some who contend the program has lost some of its appeal. They argue that it now has less appeal to farmers as farming has changed. Nonetheless, this program has enabled the state to lose far less of its farmland over the ensuing decades than adjoining states.

Following enactment, environmentalists faced the challenge of getting the new land-use planning law actually put into practice. OEC and the Sierra Club worked hard on this, as well as getting environmentally minded people appointed to the new state commission. These would ensure that that the required statewide goals and guidelines would be strong enough to protect farmlands from sprawl and urban growth (they would set the state requirements for local plans). The following year, Henry Richmond left OSPIRG, and, with McCall, founded 1000 Friends of Oregon to serve as the group that would have, as its sole purpose, watch-dogging implementation. Both John Gray and Glenn Jackson helped raise the funds to launch it. Thereafter, 1000 Friends came to play a leading role in defending it. Over time, their work formed one of the bases for Oregon land-use case law.

A flood of workshops ensued across the state to help write the goals and guidelines for the program, which the new commission then adopted in late 1974. Over 10,000 Oregonians participated in over these workshops. A byproduct of these sessions was the development of a strong cadre of supporters who could be counted on to defend the program.

And strong defenders were needed. Repeated efforts were made by rural interests to repeal or roll it back. Voters in eastern and southern Oregon were particularly hostile to it. In 1976, fifty-seven percent voted to keep the new program; in 1978, sixty-one percent voted against dropping the state's role in overseeing the program; and in 1982, fifty-five percent voted against repealing it. When Bob Straub became governor, he helped rally support to defend the program and appointed supportive people to administer it, including

Arnold Cogan, who was the first director of the program. Straub also helped convert homebuilders into being supporters in some of these contests.

By 1986, all counties, as well as cities, had completed preparation of the required plans, with all of these finally being accepted as complying with state goals and guidelines. Development of these plans with "urban growth boundaries" and restrictions on rural development proved to be very controversial. By 1993, growing pressure moved the legislature to enact some major revisions in the laws affecting rural development. In a tradeoff, they permitted some limited development on less productive lands in return for improved protection for the state's best "high-value" farmland. These changes have proved to be very effective in protecting farm and forest lands from urban sprawl. They eliminated the designation of small-scale resource lands and allowed dwellings to be built on farm and forest lands where lots had been recorded in the past for that purpose.

Further changes came in the new century, as three ballot measures emerged over the issues of "fairness" and compensation. In 2000, voters approved Measure 7, which granted compensation when regulations devalued one's property, but the Supreme Court invalidated this measure because of technical flaws in the way it had been drafted (the state's constitution does not permit amending more than one section in a measure, which this measure did).

Critics put forth a similar measure again in 2004 (Measure 37) on the issue of so-called "regulatory takings." It required compensation for any reductions in the value of one's property as a result of regulations restricting its possible use that were imposed after one assumed title to the property. The regulating body then could choose to either pay compensation (which was unlikely) or withdraw the regulation that caused the reduction. It passed, but then a court in Marion County overturned it, but then the state Supreme Court overturned the ruling of that judge. Things were getting tangled.

This compensation measure was modified further by another measure that the voters approved in 2007 that provided an alternate way to settle compensation claims. It only allowed claimants for compensation under Measure 37 to seek to build up to three homes on high-value farmlands

Source of the Metolius River

if they had been permitted when the property was first purchased. This measure, which was supported by environmentalists, did pass, and by a slightly larger margin than Measure 37. Even the Oregon Nature Conservancy joined in. This may have put an end to this line of attack for a while.

In implementing the land use law, OEC and the Sierra Club had always hoped that various environmentally vulnerable areas would be designated as Areas of Critical State Concern. This did not happen for a number of reasons. Instead of farmlands being designated as Areas of Critical State Concern (as OSPIRG urged), by default farmlands became the principal focus of the entire planning program. While farmlands and other important resource areas got their planning goals, the others were taken care of through other government programs, such as the Willamette River zone, which was covered by the Greenway program, and the Columbia Gorge, which was designated a Scenic Area. The coastal goal qualified Oregon for compliance with the Federal Coastal Zone Management Act and brought funds for this part of the state's planning program.

Chapter 5—Oregon's Environmental Laws: Key Programs

Special protection was slow in coming to one much cherished area, the Metolius Basin in the central Cascades. Although Governor McCall urged its protection in 1974, it was not given such protection under the land use program until the legislature finally did it in 2009. At that time, the basin was threatened by two new destination resorts that would have imposed almost 3000 new summer homes on it. Something had to be done to block these. With Governor Kulongoski expressing concern, the legislature at last acted.

The Metolius had long been recognized as a special place. The river flowed out of two prodigious springs near Black Butte, providing a stable flow of cold, pure waters ideal for trout, turning it into a favorite spot for fly fishing. Special designations had already been conferred on it. It was a federal wild river (1988); in the same year, the state made it a state scenic waterway; and a few years later, the Forest Service classified 86,000 acres of the Deschutes National Forest there as the Metolius Conservation Area.

Notwithstanding these designations, state planning regulations until then had allowed development on private forest inholdings. This included more extensive destination resorts. So in 2009 the legislature made the basin an Area of Critical State Concern. It provided that no destination resorts could be built in the basin, except for some small-scale, clustered recreation developments. And these must not have a negative impact on the river or its fish and wildlife.

At last the state got protection for special places that should not be spoiled by overdevelopment.

Greenways: Willamette

This program, established by action of the legislature in 1967, aims at maintaining the natural and historic features along the sides of the Willamette River on its entire 255-mile length—from Cottage Grove to its confluence with the Columbia River. Better public access is also a goal. The planning for it was confirmed by legislation adopted in 1973.

**Governors Tom McCall and Bob Straub
March 16, 1966**

While it fosters recreational use of the river, its founding documents make it clear that it contemplated continued agricultural use of the farmlands on either side of it.

The state Parks Department administers the program, provides technical assistance, exercises limited regulatory authority under it, and disburses funds to facilitate its program. It did the initial planning and coordinates efforts by units of local government, who are in charge of acquiring land, with 4000 acres acquired so far. Under its plan, much more acreage could be acquired, but acquisition has slowed down.

Three new state parks[23] have been set up within it and forty-three recreational areas have been established. More access points have been developed, as well as points accessible only by boat. It has triggered heavy use by rafters. Funding for it has come largely from the federal Land and Water Conservation Fund.

---

23 They are the Molalla River State Park near Canby, the Willamette Mission State Park north of Salem, and the Elijah Bristow State Park east of Eugene.

The Greenway has been incorporated into the statewide planning program since the outset, and its aims must be considered when projects are planned near it.

The idea for it came from state Treasurer Bob Straub, though Tom McCall thought it came from University of Oregon dean Karl Onthank, who chaired a state parks advisory committee dealing with the river. Onthank heard about the idea from Straub and then passed it along to McCall. In 1966 when they ran against each other for the governorship, McCall and Straub sparred over who could better implement the idea. When McCall won, he had to implement it, but his ideas were vague, unlike Straub, who had put together a detailed plan. McCall asked the state parks agency to put together a task force to come up with a plan. He was also reluctant to antagonize farmers owning abutting land and avoided resorting to condemnation.

But Straub had always hoped that more emphasis would be placed on public ownership along the river. When he became governor in 1975, he tried to get the legislature to at last authorize use of condemnation, but the die had been cast by the agreements made in the context of the planning program. The river zone was to be protected from urbanization but also from condemnation for parks and footpaths. In 1976, the Land Conservation and Development Commission (LCDC) established the Willamette River Greenway Goal as a yardstick for assessing the adequacy of county planning along the river. It ended the controversy over condemnation but solidified the status of the program.

Other Greenways in the Making: Frontage Along the Lower Deschutes River

Expert fly fishers have always recognized the lower dozen miles of the Deschutes River canyon as one of the best in the state. But over time, too much of its frontage fell into the hands of cattlemen, which created two problems. Access was beginning to become difficult, and cattle were trashing the streamside.

Lower Deschutes River Frontage

On top of that, they faced plans for a hydro dam there. When that fell through, there were rumors that the owner, the Eastern Oregon Land Company, planned to sell its property abutting the river to a private sports-fishing club that, again they feared, would curtail public access.

In the early 1980s, they felt they faced an emergency and went to Governor Vic Atiyeh to seek state help. He asked the director of the State Parks to do what he could to help save the area. The Parks Department thought they needed to raise $1.5 million to buy it, but their effort soon foundered.

The Oregon Wildlife Heritage Foundation stepped into the breach, launching a campaign to "Save the Deschutes." It solicited contributions across the state, with their progress charted using the graphic device of a thermometer. They poured in, from even children and the elderly, sometimes for just a dollar. In a matter of months, the sum was raised.

It turned out that the threat from a private fishing club was a phantom put forth by a land owner there who wanted a market for his property and thereby created one. In the end, he donated his land. No matter—by 1983 the money had been raised to buy out the principal owner.

CHAPTER 5—Oregon's Environmental Laws: Key Programs

Now with the frontage in state hands, the public is guaranteed access. And with grazing by cattle ended, the willows and alders have come back along the water, and there are trout in abundance. All of this reach of the river is now classified as both a federal wild river and a state scenic waterway, with a state recreation area abutting it at its terminus.

This constitutes a new model for rescuing endangered pieces of the Oregon landscape. In a way, it is another greenway in the making. A few years later the BLM set up a similar recreation corridor along twelve miles of the Molalla River through a land exchange. It runs from the Glen Avon bridge to the Table Rock Wilderness.

COMBATTING POLLUTION: AIR, WATER, AND RECYCLING

Phasing Out Wigwam Burners

At one time, when the state's lumber industry was its prime, these conical burners sheathed in metal were everywhere in the state, numbering as high as 500. They were up to sixty feet high, and almost as wide at their base, with a wire screen at the top. They incinerated as much as three millions tons of woody waste annually. Every lumber mill had one, going night and day.

They also fouled the state's air, producing not only smoky air but a heavy concentration of particulates and heavy soot-fall. Some viewed them as the price of prosperity, but many thought they were a real nuisance.

Oregon State University engineers were asked by the legislature in 1967 to suggest ways of reducing their impact on the state's air. They put forth a few technical improvements that would help. Some were implemented.

A few years later, the legislature went all the way and in 1969 set up the state Department of Environmental Quality with authority to issue regulations to improve air quality and to enforce them.[24] And it was given staff to devise the regulations and do the work. The next year, Congress enacted

---

24 The state set up its Air Pollution Authority in 1951 (the nation's first) at the behest of the state League of Women Voters, but it made slow progress.

Wigwam Burner (Toledo, OR) March, 1959

the federal Clean Air Act, which made this a requirement; they had to issue standards to clean up the air.

To deal with the air quality challenge embodied in Wigwam burners, the department issued regulations that limited particulate emissions to .2 grains per cubic foot of gas. Most Wigwam burners were either beyond this limit or near it (depending on whether they had made any improvements).

These new air quality standards were a major factor in the closing of most of these burners during the 1970s. By 1980, they had all closed.

When the lumber industry held a press conference in the early 1970s in Salem criticizing these new standards and the push to close the burners down, Governor McCall defended the standards and told the industry they were wrong. And he was someone who ordinarily did not cross the lumber industry.

Field Burning

In the 1940s, growers of seed grass in the southern end of the Willamette Valley were grappling with the problem of how to control insects and rusts that were infesting their fields. Scientists at Oregon State University suggested that they burn grass stubble in the late summer to kill them.

This worked for them, but imposed a terrible air quality problem on those who lived in communities there, particularly in cities such as Eugene. It was so bad in the summer of 1969 that the smokiest day was called "Black Tuesday." Governor McCall then had to suspend field burning for ten days. Oregon's new Environmental Quality Commission began to collect and monitor data on this problem.

In the 1980s, as many as 250,000 acres were burned in this fashion each year. In August of 1988, the smoke along I-5 south of Albany was so bad that limited visibility caused a crash, killing seven and injuring thirty-eight. This had happened despite legislative restrictions on burning between July 4 and Labor Day, when the smoke might blow into wilderness areas (EPA gave them special protection to keep their air from being degraded). Officials immediately restricted burning along the freeway.

Legislators from Eugene pressed for a solution for decades. The American Lung Association supported them, vouching for the severity of the problem. Following this accident, they finally got the legislature to curtail the practice. In 1991, it reduced the acreage that could be burned from 250,000 acres to 65,000 acres. And in 2009, it reduced the acreage further from 65,000 acres to 17,000 acres.

Now burning is only permitted on steep terrain (as around the Silverton Hills) and in a few other exceptional cases. And in those cases, burning is only allowed when the wind will blow the smoke to the east and away from Eugene and Salem.

Burning now occurs on only five percent of the land that once was burned and only when it won't be a problem. One can only wonder why it took so long.

Banning Aerosol Sprays

In 1975, Oregon was the first state to get behind rising concerns over the effect of CFCs—as from aerosol sprays—on the earth's ozone layer, by banning them. The year before, world-renowned scientists had spotlighted this danger, and the year after, Congress gave EPA authority to control their use. In 1978, it banned use of most of them. Eventually, their use was sharply curtailed by the Montreal Protocol (1987). Oregon can be proud that it gave impetus to these actions on a broader scale.

Benzene in Oregon's Air

Oregonians living near freeways having been breathing air that at times has as much as forty times as much benzene in it as the amount allowed elsewhere by EPA. Benzene is a dangerous toxic substance that can cause cancer, particularly leukemia.

Portland residents are twice as likely to fall victim to cancer from benzene as citizens elsewhere. We have had one of the highest levels of benzene in the ambient air of anywhere in the United States. These levels are not considered to be safe by our DEQ.

Much of our oil comes from Alaska's north slope, which has high natural levels of benzene. Because our air, otherwise, is cleaner than national standards, EPA for a long time did not address our benzene problem. Under President George H. W. Bush, it was not willing to take further action until environmental groups sued it in 2005 to force it to act.

Under a court order to move forward, it was going to allow refineries to buy credits elsewhere to avoid making major benzene reductions in the northwest. Senator Wyden alleged that this leeway would permit them to turn this region into an "environmental sacrifice zone," stating, "It would allow gasoline here to remain the dirtiest in the country and pre-empt Oregon from adopting tougher limits of its own." In 2003, the state's Environmental Quality Commission adopted strong rules on toxic air pollutants, including benzene, and the effects of their cumulative mixes. However, they were not implemented with much sense of urgency.

In November of 2006, Senator Wyden held a press conference on Portland's Eastbank Esplanade proclaiming that EPA "…must set a standard that protects the people of the Pacific Northwest and not oil company profits."

He then announced that he was placing a hold on a Bush administration nominee for the position of EPA's general counsel until they changed their approach (thus forcing them to get sixty votes in the Senate to proceed with a vote).

In 2007, EPA gave way and substantially reduced the opportunity to use credits. Starting in 2011, it established a uniform ceiling on benzene emissions for the country and in the middle of the following year kicked off a process of phased reductions over time. At the end of this process in 2030, benzene levels are expected to be less than half of what they now are.

But the process will be slow. By 2017, in places in Portland, benzene levels will still be thirty times the health-based benchmark level. DEQ hopes to implement a plan for reaching that level by 2021. It is now requiring gas stations and bulk terminals to capture gas vapors and to avoid topping off gas tanks. Best practices must also be used.

While one can be grateful to Senator Wyden for securing greater protection for Oregon's air (and to groups such as the OEC which supported him), one can wonder whether more attention should have been devoted over time in Oregon to this problem. It will take a long time to get the targeted reduction, and a lot of people will have succumbed to cancer by then. And we will still have some of the dirtiest, most benzene-laden air.

Recycling E-Waste

Other contemporary issues of environmental contamination have needed to be confronted too. One of them involves what to do with modern electronic devices once they are defunct—devices such as computers, monitors, printers, television sets, cell phones, and fax machines.

If they are just discarded and put in landfills, the mercury, lead, and cadmium and other toxics in them can leach into groundwater over time (becoming leachate) and threaten the environment, including drinking water.

A new approach was taken with the enactment by the legislature in 2007 of the Oregon Electronics Recycling program, for which the Oregon League of Conservation Voters lobbied hard. It bans discarding e-waste in landfills and establishes a network of collection sites in all counties and in cities of 10,000 or more people. These recycling centers must accept all donated e-waste without charge and assure their safe recycling. In some cases, the defunct products will find their way back to the manufacturers for re-use and recycling. Those who disassemble the devices here will need to operate under permits and laws applying to hazardous waste, which are onerous.

This program is being financed by the manufacturers of electronics. Twenty-six million pounds of these discarded products were recycled in 2011.

This marks a major advance in our movement toward the German practice of requiring that problematic products be returned to their producer at the end of their life span.

It builds on the promise of our pioneering bottle bill.

Another class of bio-accumulative toxics, PBDEs, were largely banned by the legislature in 2005. They were used as flame retardants and can impair immune systems when they build-up in tissues.

Protecting Groundwater

With seventy percent of its residents drinking water derived to a degree from groundwater, this issue was supposed to have been addressed by the Oregon Groundwater Quality Protection Act of 1989. It established a broad policy to protect this resource from contamination and to restore it. Past assessments of contamination in about a third of the areas relying on groundwater showed some degree of impairment in most cases, usually by nitrates and often pesticides.

The Act required that all state agencies dealing with this resource conform their rules and programs to achieve consistency with this goal. DEQ was supposed to conduct assessments and monitor conditions, and restore resources that had been impaired. More importantly, it was supposed to take steps to prevent contamination. However, it now maintains only three groundwater management programs (in the Lower Umatilla Basin, Northern Malheur County, and the Southern Willamette Valley). It has only half the staff for this program that it had in the 1990s and has had to give up assessing and monitoring the situation and providing statewide coordination of these efforts. It now simply focuses on restoring groundwater quality in these three basins.

While this program could be important, it is not being treated that way.

Energy Issues

Renewable Energy Targets

In 2007, Oregon also took a major step towards reducing our production of greenhouse gases and safeguarding the global climate. By action of its legislature, the Renewable Portfolio Standard measure requires that twenty-five percent of the state's electrical load come from renewable sources.

This applies, particularly, to the two largest utilities in the state and phases in on a graduated schedule until it is fully met by 2025. Requirements for smaller utilities, such as public utility districts, are less stringent.

Under Oregon's Renewable Energy Act, "renewables" are those comprised of energy derived variously from wind, solar, waves, biomass, or geothermal sources. It also includes power from what are termed "new" hydro sources, but excludes that derived from efficiency upgrades to existing projects or from projects in environmentally protected areas.

What cannot count toward that goal is energy derived from coal, oil, natural gas, or from nuclear power plants.

Utilities can meet this goal by building and operating plants producing "renewables," or they can buy it from other firms, including a minor share on the open market.

There is one possible loophole: utilities are excused from meeting this standard if it would require an increase in power rates of more than four percent. But in that event, they have to make alternative compliance payments into a fund that the state will use to buy renewable power or apply to reducing demand through conservation. It remains to be seen how substantial this loophole is.

This groundbreaking measure passed handily—by a margin of 41–18 in the House, and by a margin of 20–10 in the Senate. It had bipartisan sponsorship and support from all parts of the state, including rural areas. Republican State Senator Jason Atkinson of Grants Pass was its co-sponsor. In the House, then Representative Brad Avakian from the Portland area led the way. Crucial leadership was provided by Governor Ted Kulongoski and then House Speaker Jeff Merkley, who cared deeply about climate issues.

How did this happen?

Clearly, the environmentalists supported it, including the Oregon Environmental Council, the Sierra Club, NRDC, OSPIRG, Environment Oregon, and the National Wildlife Federation, and many energy-oriented environmental groups, such as the Renewable Northwest Project and the Clean Energy Coalition. And they were coordinating their efforts through the Oregon Conservation Network.

But so did the utilities: Portland General Electric (PGE), Pacific Power, and the Eugene Water and Electric Board (EWEB). The Oregon Municipal Utilities Association did too, as well as the Citizens Utility Board, the League of Oregon Cities, and the Association of Oregon Counties. Tribes and ministers also joined in.

And it also drew support across the business spectrum—from the Oregon Business Alliance to the AFL-CIO. Environmentalists apparently were able to get business interests to come to the table to hammer out an approach because they were willing and able to mount a campaign for a ballot measure on the issue. Editorial support was widespread.

All of these groups had been part of a preparatory process in which the shape of the program was negotiated in a series of "behind-the-scenes" meetings. Various compromises were reached to bring them aboard. For instance, the utilities were allowed to recover some of their upfront costs from industrial and commercial customers, rather than residential customers. At the end, it was described as a team effort.

This process did not remove all opposition. A group known as the Industrial Customers of Northwest Utilities opposed the measure because the group felt it exposed them to the risk of higher costs and rate increases.

CHAPTER 5—Oregon's Environmental Laws: Key Programs

Windmill Farm in Eastern Oregon near Columbia Gorge

And not all of the environmentalists were pleased with the way that biomass and small hydro were treated.

But by the time negotiations were ended, an amazing level of support had been built. The challenge will be to realize its promise.

Oregon is now making real progress toward this goal. In a recent year (2012), the state was getting ten percent of its power from wind. In terms of new renewable sources of power, wind is the most significant source. The production of power from wind has surged ten-fold in the last decade. Oregon now gets almost 2500 megawatts from wind projects. Wind experts think that the state has the capacity to get 27,000 megawatts from onshore wind sources. Oregon is one of the top six states in producing wind power and provides one of the fastest growing markets for it.

The state now has five major wind farms in eastern Oregon. The largest, indeed one of the world's largest such facilities, is the Shepherd's Flat Wind Farm in Gilliam and Morrow Counties, spreading its 300 turbines across 30 square miles and producing 845 megawatts of power. Its towers are huge: as many as 300 feet tall, with 100-foot blades.

Wind farms are being brought on line so quickly that the Bonneville Power Administration (BPA) is encountering a shortage of sufficient capacity on the grid to carry this power. Also at the time this is written, it is struggling with the challenge of coordinating the timing of this power with that produced by releasing water from its dams.

And these wind towers also pose other problems—environmental ones. While they seem to be compatible with ranching, some think they are unsightly, particularly as they are built along conspicuous ridgelines to catch breezes. They are not always welcomed near scenic reserves such as at the Columbia Gorge or the Steens Mountain complex.

Moreover, they pose problems for migrating birds. Some birds have problems navigating near them when visibility is poor and where they have high turbine speeds. Bats seem to suffer from reduced air pressure around fast-moving blades. Audubon Societies want them kept out of corridors used by migrating birds and are indicating sensitive zones. Increasingly the speeds of turbines are being reduced, and towers are being spaced further apart. Also after a while, migrating ducks and geese seem to learn where these farms exist.

One can hope that the conflicts with birds will diminish. But it is also true: that all power "pollutes" (i.e., it can have environmental drawbacks as well as advantages). That is true as much for renewable power sources as for conventional ones.

Demise of Coal Projects

The future of Oregon's one existing coal-fired power plant—Boardman in eastern Oregon—is not entirely nailed down, though it presently is under orders from DEQ to end operations by 2020, rather than staying in operation until 2040 as PGE had wanted. Under a 2011 consent decree settling litigation against this plant, PGE has agreed to further reduce sulfur dioxide emissions by 2020 (with a declining cap on them). EPA is still processing a notice of violation against the plant for operating since 1998 with insufficient air pollution controls.

Boardman Coal-Fired Power Plant

Boardman, one of the dirtiest industrial facilities in Oregon, has always resisted installing state-of-the-art pollution abatement controls. It is the largest stationary source of sulfur and nitrogen oxides in the state, forming smog and acid rain and fog. And it is a significant source of greenhouse gases and dangerous mercury.

Environmental groups have not been of one mind about this compromise phase-out date, with OEC acquiescing in it and the Sierra Club pressing for an earlier date. The Friends of the Columbia Gorge also has favored a quicker phase-out because the Gorge is burdened with a heavy dose of this acidic pollution (caused in part by Boardman).

But the state has already decided to have no more of this type of polluting plants. In July of 2009, Oregon's legislature passed SB 101, which sets strict standards for new base-load power plants that limit how much CO_2 that they can emit. In terms of emissions per mega-watt hour, they cannot emit more than 1100 pounds of CO_2. It is not thought physically possible for coal-fired power plants to get their CO_2 emissions down to this level.

Utilities are not permitted to evade this requirement by buying their power, on either short-term or long-term contracts, from sources outside the state that cannot meet this standard. Nor can they extend the life span of existing coal-fired plants.

And they cannot evade the requirement by building such a plant here and then offering to offset its impact by making payments into a fund that would be used to facilitate comparable CO_2 reductions.

Exempted from this law are facilities that generate power to meet peak-load demand (i.e., for immediate use) that are fueled by natural gas or petroleum distillates. It only applies to those that meet base-load demand.

A different law from 2007 (HB 3283) applies to power plants fueled by gas. Those meeting base loads must produce CO_2 emissions seventeen percent below those from the most efficient base-load plants in the U.S.

We can rejoice that we will see no more Boardmans, and even that will be gone before long. Let us hope that its replacements are really better.

Other Energy Improvements

Oregon's legislature also passed other measures in 2009 that address the challenge of climate change.

The Low Carbon Fuel Standard legislation (HB 2186) calls for a ten percent reduction in the carbon content of fuels sold in the state for cars and trucks. That improvement is measured against the levels prevailing in 2010. The reduction must be made in a fashion that does not boost prices at the pump and will be phased in over this time, taking into account the life cycle of the fuel. DEQ is to formulate regulations to implement the rule. OEC took the lead on this improvement.

To date, DEQ has just developed a system for tracking carbon intensity, but has not moved to implement the reduction, which some view as simply a subsidy for alternative fuels. If not implemented by 2015, it is set to expire—though the legislature may extend its time. Governor Kitzhaber has indicated that he will not move ahead with the program unless the law is extended.

Improvements must also be made in energy efficiency standards for both residential and commercial buildings. Mandatory improvements must be made in building codes in the range of fifteen to twenty-five percent for commercial buildings, and in the range of ten to fifteen percent for residential buildings. These changes were required by SB 79 and went into effect in 2012. Oregon also has a program mandating energy efficiency standards for television sets, many commercial appliances, and appliances not federally regulated.

Under the Clean Cars Program, by 2016 all new cars sold in Oregon must emit thirty percent less carbon dioxide and other greenhouse gases than in 2005. This is expected to be driven by mandated federal efficiency improvements for autos.

Altogether, these energy improvements are a remarkable achievement. But implementing them will be an ongoing challenge, as will the entire issue of climate change itself.

Surface Mining Using Chemical Processing: Oregon Enacts a Path-breaking Law

A noteworthy step was also taken to protect Oregon's environment from surface mining. In the late 1980s, Oregonians became worried about the spread to this state from Nevada of a new kind of mining for gold. Unlike past gold mining here, this involved surface mining, which used cyanide to leach low-grade ore from sandy material. The leachate produced could collect in ponds where it could harm birds and wildlife. A major international mining company proposed such a development south of Vale.

This prospect triggered a major debate over the future of mining in the state, with the Oregon Environmental Council (OEC) working for safeguards governing such mining. As a result, in 1991 the legislature adopted a new law reforming the use of chemicals in surface mining here.

This law requires that such mining be done in a fashion that minimizes damage to the environment. Operators must use the best technology available and practical to comply with standards that the regulators will set. The operations must aim at zero mortality to wildlife. Chemical processing solutions must be covered or contained to keep wildlife from having any contact. The acid must be applied through the use of pads that are moistened through drip techniques. Fences at least eight-feet high must be built around the operations. In the operation, there must be no net loss of habitat.

Once the operation ends, the site must be reclaimed so that it functions as a self-sustaining ecosystem. The regulators will also determine how much of the site must be filled back up. Inspectors must approve the final reclamation. A bond must be posted at the outset to ensure that the site is reclaimed in this fashion.

The out-of-state mining industry complained that the law made standards so strict that they would not come into the state to operate, and indeed the firm that proposed to operate south of Vale never opened this mine (though mostly for economic reasons). And the law was probably enacted because there was no established mining industry here to oppose it.

But the law may become valuable at last as this is being written. A new surface mine for uranium ore is under consideration just north of the Nevada border and west of Highway 95. It would use acid to leach uranium from crushed rock to produce yellow cake (uranium oxide). Apparently, the time has come when operators are willing to operate under this law that they viewed as giving environmentalists everything they wanted.

CHAPTER 5—Oregon's Environmental Laws: Key Programs

References

This material was derived from a combination of sources: interviews, organizational newsletters, and from various online sources.

Kathryn Straton, *Oregon's Beaches: A Birthright Preserved* (Oregon State Parks and Recreation Branch, September 1977).

Brent Walth, *Fire at Eden's Gate—Tom McCall and the Oregon Story* (Oregon Historical Society Press, Portland, 1994); especially chapters 7, 9, 16, and 18.

Richard W. Judd, Christopher S. Beach, *Natural States: The Environmental Imagination in Maine, Oregon, and the Nation* (Resources for the Future, Washington, D.C., 2003).

CHAPTER 6
ENVIRONMENTAL TURNING POINTS

DEMISE OF NUCLEAR POWER PLANTS IN OREGON

Trojan Plant

The Trojan plant was Oregon's only nuclear power plant. Located on the Columbia River near Rainier, it operated for only twenty-one years. While it has been closed since 1996, its spent fuel rods are still housed on site in thirty-four thick casks perched above ground. And they remain radioactive—all 359 metric tons of them. PGE's assumptions about how soon these casks could be transported to a federal repository increasingly look dubious. They were designed to last for only forty years, but half of that time has now elapsed. And the federal government has not yet settled on where the repository will be.

The nuclear power plant idea was controversial from the outset, when it was first proposed in 1968. OEC opposed it then; others gradually joined as critics. This PGE proposal only turned into a reality after emerging from a long and contentious set of hearings held in St. Helens before the Atomic Energy Commission (AEC). OEC had intervened in the hearings, and in 1971, 400 people turned out at a hearing in Oregon on this proposal by a committee of the U.S. Senate.

Even before the AEC proceedings, the Oregon Steelheaders had gotten PGE to install a water filtration system to remove chemicals used in the cool-

Trojan Nuclear Plant, November, 1973

ing tower to control algae—thereby better protecting fish in the river. As the upshot of a lawsuit that OEC filed, it also finally got PGE to do a geologic study to search for faults at the site, but none were found.

After the facility was built in 1975, opponents began to demonstrate on site to protest the way it was pushed forward. Though many were arrested for trespassing, the courts often failed to uphold efforts to punish them.

At that time, on-site demonstrations against operating nuclear power plants were happening all over the country. Most of those protestors felt the government was not regulating nuclear power in a fair and open manner. All of the licensing agencies always held in favor of the nuclear option, regardless of evidence of risk.

By the late 1980s, the irreconcilable opponents made repeated efforts via ballot measures to force closure of the plant. In fact, they did this three times, with Lloyd Marbet of Boring leading the effort. Each time PGE had to put up more and more money to defeat them, spending millions in the process.

PGE began to grow weary of never-ending contention over this plant; not only was it costing them growing amounts of money to fight these measures, they also had to make continued retrofits to remedy defects. In the early 1990s, they found cracks in the tubing in the steam generators, which would be costly to repair. This happened shortly after PGE had just assured the public that this plant was safe.

In 1992, its board of directors had enough. This was more than they had bargained for. They voted to close it down, which it did in 1996, two decades before it was scheduled to close. Its record was not much of an endorsement. Its cooling tower was taken down in 2006.

Other Proposals for Nuclear Plants

In the 1960s, the Eugene Water and Electric Board (EWEB) was anxious to get into the nuclear "game." It had put up thirty percent of the capital for the ill-fated Trojan plant.

It thought of building one near Eugene, but nearby farmers were outraged and protested vehemently. So then EWEB shifted its focus to a site on the coast near Florence at Big Creek, but that also did not pass muster. Eugene citizens objected to it too (there was a nearby fault) and countered with a local ballot measure proposing a four-year moratorium.

On the passage of this moratorium, EWEB gave up the effort to have its own nuclear plant. A contributing factor may have been a change in control of its board; the dissidents had ousted the "old boys" who had controlled it.

Feelers were tried around the state for other places to put them: elsewhere along the Oregon coast (as at Cape Kiwanda) and in eastern Oregon (as at Pebble Springs), but all were dropped. Various factors contributed: lack of public support, falling demand, escalating costs (both construction and operating), and poor prospects for surviving expected ballot challenges.

No one seemed to want a nuclear plant in their backyard.

CHAPTER 6—Environmental Turning Points

The State Votes "No"

Some of the protestors at Trojan did pursue recourse in more conventional ways. Peter Bergel and Charles Johnson had been lobbying the legislature to slow the process down and came close. So they tried using statewide ballot measures instead, going to the voters three times.

Finally, in 1980 one of their measures met with success, being approved by fifty-two percent of the voters. This time they found a way to frame the measure in an acceptable fashion, and got backing from a wide enough coalition of groups who had doubts about the rush to build nuclear power plants.

Their measure contained two prerequisites before any more new "nukes" could be built in the state. First, there had to be an operating repository for spent waste from nuclear power plants. The federal government had been trying to get one started for years and still has not overcome the endless problems that seem to arise.

The second requirement was that approval had to be first granted by the citizens of the state by a popular vote. This has proven to be an obstacle that could not be overcome. No such approval has been sought since this time.

Public suspicion has grown for a number of reasons. Too much risk seems to be involved both at the beginning of the process, when uranium is mined (leaving radioactive tailings on site), and at the end of the process, when the spent fuel rods need to be safeguarded for a quarter of a million years.

And while the plants have been heavily engineered to protect the public from radioactivity, there seem to be endless operating problems. Over the years, the nuclear licensing authorities have had to order a series of retrofits to address them, making the power more costly than anticipated.

This simply seems to be too costly a way to boil water to produce steam to turn turbines. The utilities just do not seem to be well-suited to deal with such complexity and risk. Now we know that earthquake faults and tsunamis can wreak havoc on everything in their way, including nuclear power plants sited along coasts. All too often they have been put there to have easy access to cooling ocean water.

At any rate, Oregon now has no nuclear power plants and is not likely to get any.

Mt. Hood Freeway

Portland was in the forefront of a nationwide movement to reject problem-prone freeways in urban areas. And its rejection of the proposed Mt. Hood Freeway was its pivotal case. This battle was fought over the years 1970 to 1974.

In place of a grid of freeways, it chose to bring back light rail lines and streetcars. This was now possible under changes that had been made in the early 1970s in the rules for use of funds in the federal Highway Trust Fund, stemming from a popular campaign to "bust the trust fund."

In 1943, planner Robert Moses was brought to Portland by one of its commissioners to plan a freeway system for the city. In an eighty-nine-page study, he proposed building a grid of freeways across Portland's neighborhoods. Among those proposed would have been freeways on the east side of the Willamette along Prescott Street, one along 39th Street (to be known as the Laurelhurst Freeway), one in Sellwood, and one along S.E. Division, which would have constituted the Mt. Hood Freeway.

They were designed to connect new suburbs (that were anticipated) with downtown. It was assumed that the intervening neighborhoods were expendable and would just decay.

While some proposed freeways were built, those just mentioned were not.

All of them had powerful backing. They were proposed by the Oregon State Highway Department, with strong support from the City Council and the county government. The *Oregonian* and the Chamber of Commerce were vocal boosters, with the Federal Highway Commission lining up behind the state's plans.

Resistance began to emerge in the late 1960s in the affected neighborhoods, but the objectors were just told there was nothing they could do. The result was supposed to have been foreordained. At first, they just did not know what to do to stand in the way.

But some of them began to realize that they were really part of a nationwide uprising against objectionable freeways. They obtained a copy of a national pamphlet on how to oppose freeways, entitled "Rites of Way."

Resistance first came into focus over plans to build a freeway downtown along the Willamette River, which involved widening Harbor Drive. In the summer of 1969, protestors held picnics there to raise consciousness and to form alliances. Among those who heard their message were city commissioner Neil Goldschmidt and county commissioner Don Clark, as well as Governor Tom McCall. All were sympathetic. In due course, McCall ordered that the drive be torn up and replaced with a park. Portland was the first city in America to remove a freeway.

Goldschmidt now became a full convert to the cause of stopping the Mt. Hood Freeway. Positioning himself as a leader of the effort, he put forth a barrage of arguments: that freeways would wall off established neighborhoods and lead to their decay. That would happen because the middle class would be encouraged to flee them for the suburbs, draining off the keys to their prosperity—thus weakening neighborhoods and their schools, leading to their decay and ensuing poverty. They would also increase the pollution of air there.

After 1970, federally supported projects had to be put forth with an environmental impact statement (EIS), analyzing the expected impacts on the environment (a requirement of the new National Environmental Policy Act). This one, prepared in part by Skidmore, Owings & Merrill (SOM), contained dramatic revelations: SOM declared that the freeway would soon prove to be obsolete and would not relieve congestion. It also showed that air pollution would grow, and projects now had to meet the standards of the new federal Clean Air Act.

Opponents now filed a lawsuit arguing that this freeway would bring increased air pollution and that the proposed route was poorly chosen. In early 1974, a federal judge ruled in their favor, holding that the process by which the corridor was selected was invalid because of various procedural errors. If the proponents wanted to go forward, the designers would have to start all over again. The process had already taken over a decade.

By now, Goldschmidt had become mayor and was re-elected as an outspoken opponent of this freeway. At his prodding, Portland's City Council now voted to kill this proposed freeway by a 4–1 margin. Soon thereafter

Governor McCall told federal authorities that the state was removing this freeway from its highway plans, thus bringing down the curtain on it.[25]

Moreover, McCall told them that the city wished to use the $23 million in available federal aid to build a public rapid transit system instead. Thereafter, the city went on to use federal aid to build a spreading new system of light rail lines.

This marked a major turning point toward making better choices for the vitality of cities, air quality, and curbing climate change.

PORTLAND AIRPORT EXPANSION

At this time, Portland was also trying to build other projects. Specifically, it had convinced itself that it needed to expand its international airport to compete better with that in Seattle.

Because space at its close-in site was limited, it had decided that the only place to put a new 10,000-foot runway was in the adjoining Columbia River. It would spend $170 million dredging a prodigious quantity of sand from the river (a square mile of it) and use it as fill to build a new island in the river.

They would use federal funds for airport improvements to tackle this project. After a couple of years of planning, beginning in 1968, they were ready to ask for the funds. However, they took no notice of the new federal requirement that an EIS must be prepared (after 1970) for all such projects. They had not even thought about the impacts of their project on the riverine environment.

But local environmentalists had taken notice of their oversight. They were already marshaling the resources to block it. The OEC opposed it because of its impact on fish habitat, and Dr. Carl Petterson organized a group that could be single-minded in trying to bring it down. It was known as the Citizens Committee for the Columbia River.

---

25 Goldschmidt went on to become the federal Transportation Secretary under President Carter and later governor.

Environmental lawsuits were just beginning to be in vogue, with Marvin Durning of Seattle making a name for himself in successfully pursuing such cases. OEC arranged for him to represent them and to file a federal lawsuit, focusing on the omission of any EIS. The assigned judge quickly ruled in their favor and issued an injunction against moving forward with the project until one was prepared.

Portland's Port Commission desperately sought ways to evade this ruling. Could it build it elsewhere—perhaps south of Portland near Aurora, but this was opposed by farmers there and was too remote. Could it raise the needed money without federal aid? Not really. Could it get environmentalists to back off? An old contact of mine from my Harvard days (attorney Borden Beck) who represented the commission phoned me in San Francisco to see if we could be satisfied in some other way? No, I would not help. Could they begin dredging while the EIS was being written? Not allowed.

Reluctantly, they finally came to grips with this requirement and wrote the EIS, which revealed more damage than they imagined. The question now was could they go ahead in light of this massive amount of adverse impact? At last they concluded that they could never get clearance to proceed from the federal government.

In 1973, they dropped the expansion project and proceeded to upgrade the existing runways.

Environmentalists in Oregon now understood the power of this new tool in their hands. Through the courts, they got the same answer as with the Mt. Hood freeway. Where project planners intended to rely on a federal partner, they needed to come up with better projects. They were beginning to understand that environmentalists now had clout.

Nestucca Spit Freeway

The State Highway Department faced opposition to proposed freeways elsewhere too, especially along the Oregon coast. One skirmish unearthed the revelation that freeways were planned all along the coast. That skirmish was over their plans to build a freeway down the three-mile long Nestucca Spit near Pacific City. It was just south of scenic Cape Kiwanda—from whose beaches dories were launched into the surf (the only such place).

The contest over this proposed freeway drew heavy opposition because many regarded it as a treasured place. In fact, twenty of the opponents then had beach houses there. Part of it was now a state park too. Visitors came there from all over the state to enjoy it. They felt that freeways and their traffic did not belong in parks; they should be kept natural.

They formed a local group to mount opposition: the Citizens Committee to Save Our Sands. One of them, the Buxton family, even filed a lawsuit to stop the freeway, claiming that the land had been granted to the state just for a park, with a reversionary clause if the land was used for any other purpose, such as a freeway. In that event, the land would go back to the BLM.

While there was also local support among tourist promoters, opposition began to grow around the state. In light of it, the Highway Commission's chair, Glenn Jackson, began to backtrack. He now characterized it as a "scenic highway," which would only be two lanes, not a four-lane freeway. However, opponents were still dubious, since the bridge that was planned across the Nestucca River at the end of the spit was still wide enough for four lanes.

In 1966, the local opponents managed to get State Treasurer Bob Straub, to embrace their cause, as did the OEC. Straub called for public hearings on the proposal, and that spring led a protest hike there. Many opponents showed up with their families and marched to demonstrate their opposition, but they faced an angry group of proponents who brandished guns, threatening them with being charged with trespassing over what they claimed were privately owned dry sands north of the park.

With the state refusing to relent and to hold hearings at that point, Straub decided to hold his own unofficial hearings to provide a place where public feelings could be voiced. He did this in August of 1967 at Portland State University—where 700 people showed up to speak, almost all in opposition. Among them were university biologists who affirmed the value of this site as habitat for biota; they said it would be a tragedy to to lose any of it.

At that point, the statewide opponents shaped themselves into the Committee to Save the Beaches, which released the results of a poll showing that eighty-six percent of the state's citizens now opposed this freeway.

Now that the issue of BLM's role had been raised (owing to its reverter clause), Governor McCall wrote to Interior Secretary Stewart Udall urging that he clear the way for the freeway to be built. On learning this, Straub flew to Washington to meet with Udall and urged that he not do that but instead oppose the freeway, which Udall subsequently did. Glenn Jackson then directed his engineers to find a route that did not touch the parcels that had come from the BLM.

Straub's group argued forcefully that these beaches belonged to all of the people of the state and not to just those in the coastal counties. To counter the weight of the 700 people who showed up at Straub's Portland hearing, local supporters got 700 people to sign a petition in support of the freeway.

McCall and Straub kept arguing in public about this issue, with McCall now cautioning that this issue was not yet decided. (It was later revealed that McCall had reached an agreement to support Jackson for personal reasons.)

Straub argued that the public had not been told that there were really three alternative routes, and that the higher routes inland were really better. They were more scenic, cheaper to build, and easier to maintain. But many in the media were skeptical about Straub's motives, seeing him at this time as a political opportunist.

By the latter part of 1967, the antagonists were having dueling hearings. On one day in Portland at Benson High School, the opponents put on another of their own hearings, with 600 showing up to vent their opposition. They now unveiled a petition with 12,000 signing to oppose the project.

The next day in Tillamook, the state Highway Commission finally held an official hearing, but it was overwhelmed by immense crowds. It took two days to hear all of them, with the sessions going until ten o'clock at night. These were boisterous and contentious. Coastal commercial interests showed solidarity in pressing for the freeway, while opponents from around the state showed up in numbers as well. By the end, it was almost a draw, but opponents did have a 3–2 edge.

At this time, the state now acknowledged that alternatives were being considered: in fact, four of them. And the Oregon Coast Association made it clear that they were looking forward to the highway engineers using spits all along the coast as convenient routes for new freeways.

The controversy now morphed into the broader question of the public's right of access to the beaches over the dry sands. That story has already been told. But at the end of two years, the Highway Commission had to contend with a basic, new fact that changed everything. In December of 1969, the Oregon Supreme Court ruled that the state held in trust the public's right to cross the dry sands to reach the wet sands, and subsequently the state even designated its beaches as a state recreational area.

On learning of those decisions, the state Highway Commission decided that it would no longer contemplate using the Nestucca Spit for routing its relocated Highway 101. After almost triggering another routing controversy, the commission withdrew those plans and announced that it merely intended to improve the highway in its present location, as it eventually did all along the route.

In 1972 Governor McCall brought the whole routing controversy to what turned out to be a final end by declaring a moratorium on beach routing until a comprehensive study was done and an overall plan adopted. Lacking those, the moratorium proved to be permanent.

While ostensibly the fracas over Nestucca Spit was just about it, it proved to be a test of wills over how the coast would be treated by the Highway Commission. Having had enough in tangling with environmentalists, it elected to stick with the route of the existing highway.

In due course, the state renamed its state park on Nestucca Spit as the Robert W. Straub State Park. He may have done more for the beaches than anyone since Os West, who also has a park named after him forty miles to the north. It is too bad his sincerity was doubted at the time.

Protecting Eugene's Wetlands

For many years, these low-lying wetlands areas at the edge of Eugene, out West 11th, were not seen as having any value other than for development. Until developed, they were often seen as unsightly, with debris collecting on them.

And Eugene was anxious to attract new industry to the area, being unsure of the future of its once robust lumber industry. As it was developing its first comprehensive plan in the early 1970s, the chair of the Lane County Audubon Society's conservation committee, Sydney Herbert, believed that not all of their undeveloped tracts should be zoned for development. Some, she thought, should be kept as open space and made into parks and other reserves.

She decided she would invite the public to check out the qualities of about half of them. On summer evenings, she conducted tours around to inspect them, which included botanists, such as Stan Cook and David Wagner, and their students from the nearby university. When these experts looked at the tracts along Willow Creek, they realized that these were prime examples of the prairie terrain that once typified the Willamette Valley. And since the tracts had never been plowed, they were now rare. Less than one half of one percent of this habitat remains.

On this now scarce habitat, they discovered all sorts of rare and unusual plants. And what is more, the botanists even found one that was thought to have become extinct. This was amazing; these were very rich sites.

But efforts to get them protected in the 1970s did not go anywhere. These particular tracts were owned by local attorneys who wanted to develop them. If thwarted, they said they would plow them and plant Christmas trees. But their local state representative at the time, Ted Kulongoski, warned them that they would destroy their own reputation in Eugene if they destroyed the biological value of these tracts.

And at that time, The Nature Conservancy did not show much interest in these areas, having committed itself to other priorities in the state. By a decade later, it had reconsidered—deciding now that this tract deserved protection. It began by setting up its Willow Creek Preserve, initially leasing

Eugene Wetlands with Exhibit Map

508 acres. It then cleaned up its leased land and introduced summer fire as a management tool. And it welcomed winter flooding since the native flora thrived on it, as well in response to being burned.

But through most of the 1980s, the tug-of-war continued. The city was strongly committed to fostering an electronics industry in the area, having invested $12 million in building accommodating infrastructure. And it had begun to make progress, with a cluster of such firms now there. But the continuing conflict over the future of the area also created uncertainty about what the future would bring. This hindered further development.

For a while, guiding federal policy regarding wetlands was also enveloped in conflict, but in 1987 it resolved its conflicts under new laws and a new policy: the policy of "No Net Loss." Owners were required to preserve wetlands on their property, or replace them somewhere else. And the regulations governing replacement were long and complicated. Eugene thereupon became worried that investors would lose heart and locate elsewhere.

What was worse, they now realized that many of these tracts would fall under this new policy. When a city biologist did an inventory of natural resources there, he came to the startling conclusion that there were 1500 acres there that the federal government would now classify as wetlands. The city had not thought of them as wetlands because water did not stand on them year round—being flooded only in the winter. But the definition now turned on whether they had soils typical of periodically flooded land. And they did.

Once this new legal regime came into effect, electronics firms that were already there struggled to adapt. A manufacturer of bar-code readers, Spectra Physics, had enjoyed success and wanted to expand onto adjoining wetlands. It spent over $900,000 developing thirty acres of replacement wetlands, but the process of doing so was full of frustration, delay, and uncertainty.

The city had a bad feeling about the outlook. In 1988, it asked Steve Gordon of the Lane Council of Governments to search for a way out. He wanted to pull all of the stakeholders together in a collaborative process. He began by putting together an expert team that worked with all of them: the 125 affected land owners, industrialists, environmentalists, and the public to help them see the realities. The first thing he tried to get across was that the new wetland policy came from the federal government, not the city.

In the wake of many hearings and workshops, gradually greater trust began to take hold among the parties. And they got a better grip on the facts: the city prepared an inventory of the wetlands, mapping and evaluating them in detail; EPA put up about half of the money needed to do this. And The Nature Conservancy brought in a fresh face for the process.

Out of their dialogue, the notion that emerged took the form of seeking a multi-pronged solution: one with multiple objectives, involving both preservation and development. It would be best, they thought, to strike a balance, in which everyone would feel they got something. This would be a consensus solution that the city could adopt.

Their plan protected the more valuable sites: 1278 acres of them, and allowed development on the less valuable ones: 259 acres. Instead of forcing those who wanted to develop these to build a replacement wetland, they could now opt to pay into a mitigation bank. The bank would guarantee that the replacement wetlands would be built—contracting with various entities to build them (with the BLM, the city, or The Nature Conservancy).

The plan grew out of the city's inventory, which provided the factual basis for it. It was known as the West Eugene Wetlands Plan and in 1992 was adopted by the city, and subsequently by the state and federal governments. In due course, the plan became a model for how other cities could deal with similar problems.

Over three million dollars now have been obtained from the federal government to implement it, with others investing as well. The congressional delegation lent its help to secure these funds.

Now the plan has grown. Over 3000 acres are protected, providing rare wetland habitat for 350 plants and animals, including three threatened plants and one endangered creature—the Fender's Blue Butterfly. Beaver and river otter have returned, as well as frogs and an abundance of varied dragonflies.

A number of agencies and organizations are involved in acquiring and managing land there: the city, the county, the BLM and The Nature Conservancy. They view the project as a partnership.

And it is flourishing and continues to take form, with interpretive signs and roadside viewing places. In time, more land will be acquired, it will become more robust, and its biological diversity will grow.

While this project stands as an example of what can be done under a collaborative process, it worked so well because of its context. The parties were deadlocked; they were feeling the pressure of a new federal requirement; the city was committed; the stakeholders chose to participate; and there was a source of money to fund it.

Moreover, the person who guided it, Steve Gordon, was competent and talented. He knew what he was doing.

This was not the only important wetland in the Willamette Valley to emerge from the struggle to gain protection. Protection came to the 147-acre Jackson-Frazier wetland just north of Corvallis in the late 1980s as the upshot of a tenacious campaign led by Bob and Liz Frenkel of Corvallis. This habitat featured two rare species and supported over 300 species of flowering plants.

Under the state's land use program, it was first inventoried by the county in 1982 and identified as needing protection. But efforts to save it were frustrated because the local voters would not fund its acquisition. The Frenkels then would not accept various efforts by the county to provide lesser protection. They and their supporters brought intense pressure on the State Land Use Commission to reject a series of county efforts to secure protection under its goals for natural resource planning, and then environmentalists went to court to challenge a limited approach under federal and state wetlands regulations, and, led by the Frenkels, prevailed. Finally, through a tax foreclosure, the property fell into the hands of the county, which in 1992 made it a county Natural Area. Along with other Benton County Natural Areas, it then came into compliance with the state's land use program. It is also identified in the state's Natural Heritage Bank as a Natural Area.

While there are a scattering of other small natural areas throughout the valley, these two reserves in Eugene and Corvallis stand out. Harboring important values, they were protected by the determined efforts of citizens.

Herbicide Spraying

While those arranging solutions at places such as the Eugene wetlands knew what they were doing, those who were spraying herbicides on forests west and northwest of Eugene in the 1970s did not know what they were doing. They thought they were simply curbing the growth of brush and thereby fostering the growth of the forests they wanted. The herbicides were sprayed from light planes and helicopters.

Though there were prescribed setbacks from inhabited areas and bodies of water, such as streams and lakes, the herbicides often drifted into those off-limits areas. The spraying was primarily being done at the behest of officials of the Siuslaw National Forest, then the leading producer of timber.

But the real question was the safety of the herbicides when they drifted onto non-target biota. A number of herbicides were being used, but they included 2,4,5-T, which was already controversial for its use in the Viet Nam war where it was used as a defoliant and known as Agent Orange (combined with 2,4-D). In due course, it was shown to include traces of dioxin, which causes cancer and birth defects. It is among the most the deadly toxins; there is no safe level of exposure to it.

This was an explosive controversy because many professionals from Eugene lived on inholdings intermingled among these public lands. These professionals included doctors, teachers, and other professionals who were well educated, were used to being listened to, and who had the resources to get other professionals to help them, such as lawyers.

They began to notice impacts on their property: their landscaping was shriveling and their livestock were bearing deformed offspring. And most ominously, they began to compare notes with their neighbors and noticed an unusually high rate of miscarriages among them.

They soon got together to form a group to curb the spraying: Citizens Against Toxic Sprays (CATS), led by Carol Van Strum (whose book, a *Bitter Fog*, the Sierra Club published). Another activist, Jean Anderson from Swisshome, even got a court order preventing spraying near her ranch in 1973. The OEC soon backed their cause.

Their efforts to get enlightening information from the Forest Service were initially rebuffed. At first, they could not even get copies of the EIS. Other agencies also were not forthcoming.

So in May of 1976 they filed suit in federal court charging that the EIS the Forest Service had prepared was inadequate because it failed to deal with the kind of problems they were enduring. Soon the judge ruled in their favor, ordering the preparation of a new EIS, which dealt forthrightly with the dangers of the herbicides being used. While this was being written, he stopped use of these herbicides on the Siuslaw National Forest. This was the beginning of a see-saw process that continued for a number of years, with victories that were often short-lived and quickly countered.

In July of 1977 the *New Yorker* ran a serious article by Thomas Whiteside on the deadly nature of this herbicide containing dioxin, culminating with the question of why EPA was failing to act in light of worldwide information about its terrible toll.

Soon the regional forester temporarily suspended the use of these herbicides throughout the region. And the newly written EIS showed a high correlation between use of these herbicides and miscarriages. Notwithstanding this key revelation, the judge then allowed resumption of the spraying, though he cautioned that they must follow the directions for their use, which supposedly would protect them. Ninety thousand acres were soon sprayed with 2,4,5-T (1978).

After it also prepared an EIS that took into account this new information, the region imposed some new controls and required monitoring of operations, which resumed elsewhere.

At this time, national columnist Jack Anderson carried a story from Oregon reporting miscarriages in this area were running at a level three times higher than normal.

And others with responsibilities got into the act. Then Assistant Secretary of Agriculture Rupert Cutler prohibited spraying within one mile of permanent homes and within a quarter of a mile of streams. He tried to assume personal control of further Forest Service spraying. And Forest Service Chief McGuire imposed additional conditions on the use of these herbicides: they could only be used when there were no other options, and where cost was not the driving factor.

Then Oregon Governor Bob Straub also issued directions to his state forestry officials to tighten their controls over spraying on private forest lands. First, notice had to be given to affected property owners; printed notices also had to be posted around the area to be sprayed; and a 200-foot buffer had to be established around homes, roads, and waterways.

But Cutler faced stiff protests from Oregon's timber industry. It said his restrictions would "make a mockery of reforestation" in Oregon. Even Governor Straub objected to his approach, saying it went too far. As a result, Cutler relented, giving in to Straub—with his approach applied everywhere.

Just as everyone was adjusting to this approach, EPA at last intervened in a decisive way. It had been studying use of this herbicide, particularly around the town of Alsea. Its study was triggered by data submitted by Alsea resident Bonnie Hill, who found an unusually high number of miscarriages among women she knew and correlated them with spraying operations. EPA found convincing evidence of this correlation and announced that it was banning the use of 2,4,5-T for this and all uses everywhere. It said changes had to be made. Its announcement drew national coverage by the media.

But Oregon's timber industry was vehement in denouncing this decision, with support from forest scientists at OSU. Only later was it discovered that they were paid consultants of the producer of Agent Orange: DOW chemical.

EPA's decision unleashed a barrage of lawsuits throughout the region to curb the use of these herbicides: on other Forest Service lands, on BLM lands, and on private lands. In 1984, a federal court stopped the Forest Service from using this herbicide throughout the region (Oregon and Washington). In another case, a National Environmental Policy Act (NEPA) violation led the judge to issue an injunction stopping all spraying with this herbicide, both on the ground as well as from the air.

By 1983, EPA had worked out an arrangement with the DOW chemical company to withdraw 2,4,5-T from the market entirely. Thereafter, the Forest Service began to gradually phase out aerial spraying—even with less problematic herbicides. It found it could manage without them, thereby avoiding conflict with citizen groups such as CATS. It should be noted, though, that the Siuslaw forest was slow in making the transition, because it grew brush prolifically.

Those who forced change in herbicide use in the hills near Eugene did the whole country a service. While they had the provocation to protest, they also had the confidence, skills, and determination to bring events to a head in a productive way. We owe them for making a difference.

Demise of the Elk Creek Dam

Blocking the construction of ill-advised dams was part of the story of an earlier era. In more recent environmental times, the story-line shifts toward the removal of dams that are now are regarded as mistakes—ones made in an earlier era.

Dams on the Rogue River are particularly seen in that light. This is because the Rogue was once among the most productive salmon rivers; its productivity had been sharply reduced by these dams. And with the catastrophic collapse of salmon numbers throughout the Northwest, many feel it is urgent to begin restoring it.

Gradually low-dams on the middle Rogue have been removed as they became obsolete: Savage Rapids, Gold Ray, and Gold Hill Diversion. Their owners chose not to invest in updating facilities for the passage of fish. Had they sought to have the dams re-licensed, they would have been required to make these investments. It no longer made economic sense.

For the first time in a century, the Rogue runs free to the sea—for 157 miles.

But this is not the case in its upper reaches. In 1962, Congress authorized two dams there: the Lost Creek dam, which was finished; and the Elk Creek dam, which was not.[26] It was five miles downstream from the Lost Creek dam. The Lost Creek dam had been less controversial, with the state Izaak Walton League supporting it in the hope that it would enhance summer flows and encourage Chinook Salmon.

But ever since the 1970s, the Elk Creek dam has been a lightning rod for controversy. Senator Hatfield always pushed it, but environmentalists have fought it unrelentingly. When $114 million had been spent on it and it was one-third finished, Oregon Wild sued in the 1980s to stop it. Finally, in 1987 the Court of Appeals ruled in its favor and enjoined further construction.

In the 1990s, agencies voiced concerns: the Forest Service and the BLM said the dam unreasonably diminishes habitat for wild anadromous fish in

---

26 At the same time, Congress also authorized a dam on the Applegate River, which was finished. However, it is not on the main stem.

Elk Creek Dam and Notching Explosion

Chapter 6—Environmental Turning Points

parts of the downstream river that are classified as a wild and scenic river. The Fish and Wildlife Service went further—calling for complete removal of the dam. Even the Corps of Engineers admitted to doubts about whether completion of the dam was advisable.

Finally in 1995, the Corps of Engineers abandoned efforts to complete the dam, saying it faced serious legal and fiscal obstacles. Two years later, it asked for authority to cut a notch in the partly completed dam to allow fish to pass through it. But Representative Greg Walden, who represented the area, balked and persisted in seeking funds to complete the dam.

The problems compounded in 2000 when six environmental groups filed a new lawsuit against the Corps of Engineers to complicate their efforts, alleging violations of the Endangered Species Act and failure to consult properly with the National Marine Fisheries Service (NMFS) on impacts on a threatened salmon species. NMFS has jurisdiction over the marine leg of the migration of anadromous species.

While this suit was still pending, at last a compromise was worked out in 2008. The Corps of Engineers was allowed to remove about fifteen percent of the dam to create the notch, which was done using explosives.

Now they are hard at work restoring the area, rehabilitating 2600 acres by planting willow trees. Inventories have been done of the plants and animals that are now moving in. All of this is happening where the pool had been planned. Over two million dollars have been spent so far on restoration.

And the fish are finding it possible once again to make their way past this place and along nearby Elk Creek. But no one knows what to do about the Lost Creek dam. It is the last big barrier.

But the progress in restoring fish runs on the Rogue is impressive.

Portland's Big Pipes

Many thought that the job of keeping sewage out of the Willamette River would have been accomplished decades ago when most cities built sewage treatment plants. But one major challenge remained unaddressed for years afterwards—the job of dealing with Portland's system of combined sewage and storm-water pipes. About seventy percent of its system consisted of these combined pipes.

Every time it rained hard, the treatment plants were overwhelmed by the volume of water, and the untreated sewage was released directly into the Willamette River through forty-two outfalls (another twelve would go into the Columbia Slough). This would happen between fifty and eighty times a year. The fecal matter and harmful bacteria in these overflows posed a real risk to human health since the river is used for boating and recreation.

Since the early 1970s, Portland had been legally obliged to end this practice. Under the Clean Water Act, the Safe Drinking Water Act, and the Endangered Species Act, the state Department of Environmental Quality (DEQ) had been obliged to get Portland into compliance under permits it had given Portland for these outfalls.

Decades passed without the problem being cured. There were similar problems in other cities across the country. Cost was the obstacle: some estimated it would cost them up to $100 billion nationally to re-engineer their pipes. It was felt that it might cost Portland over a billion dollars to do it. No one wanted to bear costs of this magnitude. Cheaper ways were sought to mitigate the effect.

But environmentalists were getting restless with the foot-dragging. They were beginning to file lawsuits to force action. In this fashion, they forced San Francisco to build a deep vault underground near the ocean to hold its contaminated overflows.

When Portland's five-year discharge permit was about to expire in 1989, DEQ allowed it to remain in effect until the city would consent to what DEQ wanted. DEQ was asking the city to meet federal water quality standards at all its discharge points, including at its ones for the combined sewers. Portland complained again that it needed more time and could not clean up the problem in the next five years.

At this point, a gutsy group of lawyers in Northwest Environmental Advocates, headed by Nina Bell, filed a major lawsuit in federal court on April 16, 1991, challenging the city's practices. While it eventually did not prevail in its contentions, the pendency of the suit forced the city to come to grips with the reality that time was running out for "dodging and weaving" to evade these legal obligations. On various grounds, environmentalists were ready to go to court in a serious way.

A few months later (in August of 1991), the city came to terms with DEQ. The city agreed to replace the antiquated combined sewage system within twenty years, and as a reward got its interim discharge permits for these contaminated combined-sewer-overflow (CSO) outfalls. The completed system keeps six billion gallons of raw sewage out of the Willamette River.

To handle the volume of runoff at rainy times, the city has designed a series of huge pipes 120 feet below the surface. At last in 2003, it began construction, with construction completed in 2011. This was the biggest public works project in the city's history and cost $1.2 billion, which was financed by bonds to be paid off by city rate-payers.

Three huge pipes have been built underground: the Columbia Slough Big Pipe, which is twelve feet in diameter; the West Side Big Pipe, which is fourteen feet in diameter; and the East Side Big Pipe, which is the largest—at twenty-two feet in diameter. Huge underground boring machines dug tunnels for them. The big pipes collect the surge of rainwater and sewage and hold it until it can be pumped to the treatment plant on Columbia Boulevard. But the passage through this treatment plant only does half the job; it does not remove toxics, which are dumped in the Columbia River, where they harm fish.

Altogether these big pipes dramatically reduce the overflows from Portland's combined sewer pipes—by ninety-four percent. Instead of overflows occurring fifty times a winter, they are expected to occur only about four times.

These big pipes will remove one of the most significant remaining sources of sewage that pollutes the Willamette River. It was a long time in coming, with Portland just making its twenty-year deadline.

And Portland's water users will be bearing a heavy financial burden. But that is the cost of living in peace with the environment. The fish will bear less of a burden, and people will enjoy the dividend of living in a healthier environment.

REFERENCES

William G. Robbins, *Landscapes of Conflict* (Seattle, University of Washington Press, 2004).

Much of the material in this chapter was drawn from information derived from interviews, written recollections in my files, and from online information.

CHAPTER 7
THE ADVENT OF NEW RESERVES

Oregon Dunes National Recreation Area

In the mid-1950s, the National Park Service (NPS) did a nationwide survey of coastal resources that might merit designation as National Seashores. When the report on the Pacific coast was released in 1959, Oregon learned that one of them was along its seashore. It was the Oregon Dunes, located between the Siuslaw River and Coos Bay.

Its dunes are spectacular. These are the country's most extensive coastal dunes, with the tallest Barkan Dunes, which are the heavily sculptured ones, rising 500 feet above the sea. They are among the most impressive dunes in any location, and the Park Service found them to be of national importance.

They run for fifty miles along this coastal stretch just west of Highway 101. Summer winds from the northwest blow ocean sand inland, where it dams many local streams, creating lakes and ponds; winter winds do likewise but from the southwest.

Abutting vegetation wages a never-ending struggle against the invading dunes, sometimes over-topping them and sometimes being buried. Hills back of the dunes often are comprised of dunes overcome by vegetation that won the contest. European beach grass, introduced to protect the highway, now is conquering too much sand along the shore, including habitat for the endangered Western Snowy Plover.

Oregon Dunes

Soon after learning of the NPS report, Senator Richard Neuberger introduced a bill to make this area a National Seashore under the National Park Service. Following the Park Service findings, he included the Sea Lion Caves a bit farther to the north near Cape Perpetua, as well as some of the nearby lakes (Woahink and Siltcoos) to the east of Highway 101. He also included Honeyman State Park, which had benefited so much from CCC work.

But his bill unleashed a torrent of local opposition. The reasons varied: some is was from those who rode dunes buggies there, which they feared would not be permitted by the Park Service; some of it was from those who preferred Forest Service management for these were parts of the Siuslaw National Forest; other opponents were the owners of the Sea Lion Caves site who did not want to lose it, nor did the state want to lose Honeyman park. Others were simply suspicious of a sponsor from an urban area.

Some were also owners of summer homes around the lakes who did not want their homes to be in the new seashore. Among them were lakeside home owners who were on the faculty of the University of Oregon and probably knew Senator Wayne Morse when he was a colleague there. At any rate, they enlisted him to oppose the bill.

CHAPTER 7—The Advent of New Reserves

Robert Duncan, who became a congressman from southwest Oregon (1963–67), and later from the Portland area (1974–80), introduced a companion bill to Neuberger's in 1963. But Duncan's bill was smaller (for 30,000 acres rather than 42,000 acres) and did not include the land around the two lakes (Siltcoos and Woahink).

While Duncan was never able to get his bill out of committee in the House, he did influence preliminary action in the Senate subcommittee, which shrunk the Neuberger bill. Richard Neuberger died while in office in 1960. His wife soon succeeded him and continued to champion his bill. But the bill could not get further because Morse did not want it passed, threatening a filibuster. A deadlock ensued.

Local field hearings for the Senate had been held in the spring of 1963 in Eugene (earlier House hearings had been held in Reedsport and Florence). In my job for the Sierra Club, I had been supporting Neuberger's bill. In preparing for the hearings, I decided to go to the Park Service's regional office in San Francisco to see what the Park Service planned to unveil at the hearings. I had heard that they planned to present their proposed plans for developing the area.

To my consternation, I learned that they planned to pave a new road down the beach south from the Siuslaw River. This area was then a wild beach. My clients (the Sierra Club, the Federation of Western Outdoor Clubs, et al.) strongly opposed this, and I did too. I argued in vain that this was not only a bad idea, but a tactical blunder. It would alienate their strongest supporters. But they would not budge.

At that point, we lost interest and ceased to promote the Park Service bill. Thereafter, it became a standoff between Morse and Maurine Neuberger. Nothing happened until both of them left: Neuberger in 1967 and Morse in 1969.

Finally, after Morse had left and Hatfield became the state's senior senator, Hatfield tidied up the matter by pushing through legislation in 1972 to make the area a National Recreation Area instead, keeping it under the Forest Service and omitting the three controversial areas (the Sea Lion Caves,

Honeyman State Park, and the lakes east of the highway),[27] but going farther south to Coos Bay. By then, John Dellenback represented the district in the House (1967–75) and carried the legislation there. The objectionable road to the south was never built.

In 1972, 31,000 acres of these dunes got better protection and recognition under this designation. It is enjoyed by a million and a half people each year.

Hells Canyon National Recreation Area

For a number of decades, rivals contended to build dams in the nation's deepest canyon—Hells Canyon along Oregon's border with Idaho. Supporters of public power wanted to build a high dam, and supporters of private power favored a lower dam, both just above the Snake River's confluence with the Salmon River, flowing in from central Idaho.[28]

After the Federal Power Commission (FPC) issued a license to the private company, the issue became enmeshed in litigation that eventually reached the Supreme Court. In its ruling in 1967, the Supreme Court remanded the matter back to the Federal Power Commission, holding that the FPC also had to consider the option of not building either dam there, but keeping that part of the river untamed. Justice William O. Douglas from the Northwest wrote that opinion.

When the FPC re-opened its administrative proceedings on the broader set of options, the Northwest representative of the Sierra Club, Brock Evans, intervened on behalf of the "no dam" option. The dam proponents fiercely opposed his admission as a recognized intervener, but he could hardly be turned down in light of the court's ruling. He prevented a quick ruling in favor of a dam, with this proceeding staggering on for years (until 1972).

---

27 In 1975 Lilly Lake just south of the Sea Lion Caves was acquired owing to the efforts of Senator Hatfield and Governor Straub, in response to the urging of the state Sierra Club, led by Ronald Eber.

28 Two smaller dams had been already built upstream in the canyon by a private utility.

Hells Canyon from Hat Point

By the late 1960s, environmentalists seized the initiative and started promoting legislation offering a comprehensive alternative. It was to establish a Hells Canyon National Recreation Area[29] that not only ruled out any further dams along the middle Snake, but designated 700,000 acres in the canyon for non-commercial purposes, emphasizing retention of its natural character. Brock Evans of the Sierra Club and Larry Williams of the OEC took the lead in putting this proposal together, along with the Hells Canyon Preservation Council, which Brock helped put together. They persuaded Oregon's Senator Packwood to sponsor it, though this drew criticism from the *Oregonian* for his daring to do this without "consulting" the rest of the delegation. They also got Representative John Saylor of Pennsylvania to sponsor it in the House.

As part of their efforts to promote it, they persuaded various celebrities such as Arthur Godfrey (1969) and Pete Seeger (1972) to join in a well-publicized float trips. Godfrey also wrote letters of support and testified at Washington, D.C., hearings.

---

29 The canyon had been a Forest Service Limited Area, which was a distinctive designation in the Northwest for areas that needed more study of their wild character.

But for a number of years, few of the other members of the congressional delegations from Oregon or Idaho would declare themselves on moving in this direction. The trouble was that most of them had been committed participants in the struggle between public and private power in the canyon. Representative Ullman, who then represented the eastern Oregon district, had been a vocal champion of the high Hells Canyon dam, a public power project. And it was believed that Hatfield leaned toward the smaller private power dam.

Finally, to move things off dead center, Larry Williams put out a press release in Portland asserting that Hatfield supported putting a dam in Hells Canyon (not specifying which type), and criticizing this stance. Hatfield responded tartly that this was not the case. For the first time, he made it clear that he did support preserving Hells Canyon, but it was not apparent how.

That exchange "smoked out" Ullman too. He then joined Hatfield in sponsoring a bill that they called a "Parklands" bill. But this bill was not all it seemed. While it did block a dam in the canyon, it had no limitations on what the Forest Service could do. Grazing, logging, and road building were unhindered; the Forest Service would have had unlimited discretion.

Williams, and Doug Scott, who had now succeeded Evans, soon discovered the reason. The Forest Service had drafted this bill. While conservationists did appreciate the fact that Oregon's delegation was now willing to part company with dams there, they wanted more—real protection for this unique landscape. They then prepared a detailed analysis of each section of the bill, showing just how bad it was. They declared this bill was completely unacceptable.

Hatfield and Ullman resented this open criticism, but they eventually came around and proposed a much more balanced proposal. This new bill had meaningful provisions to protect wilderness, fish and wildlife habitat, and unique ecosystems. Portions of the river were added to the wild rivers system, and portions of the adjoining land were designated as wilderness, with further study of additional sections. At this point, Packwood embraced it. The Hells Canyon Preservation Council then lined up support in Idaho, with Senator Frank Church now supporting it.

CHAPTER 7—The Advent of New Reserves

In 1975, Congress established a National Recreation Area there of 652,000 acres. It was dedicated by an auspicious assembly of officials on July 31, 1976. A visitor center near Enterprise now tells the story, celebrating the struggle to save the canyon and one hundred miles of its river, which continues to flow freely.

John Day Fossil Beds National Monument

This national monument in the center of eastern Oregon stemmed from quiet action behind the scenes. The area had been a state park, but the state lacked the resources to do much to properly interpret it.

Citizens of Grant County, which includes some of its units, wanted to do something to improve its economy by bringing more visitors to the area. Led by Gordon Glass, in the 1970s they approached the director of its state parks, Dave Talbot, seeking ways to improve interpretation. As an upshot of their discussions, they agreed that the National Park Service could do a better job.

With the aim of involving the NPS, they approached their congressman in the 1970s, Representative Al Ullman. He agreed to help and arranged for the Park Service to evaluate the sites for their suitability as a national monument.

After making such an evaluation, the Park Service agreed that the fossils in this area were of national scientific significance and merited status as a national monument, even though the recreational features were limited.

Only a limited amount of exploration has been done there to uncover fossils. The deposits there display a record of fossil plants and animals over most of the Cenozoic Era—a period of 40 million years. This is a deposit showing evidence of one of the longest geological records. An impressive variety of ancient animal fossils have been found, among them rhinos, camels, horses, giant dogs, pronghorns, peccaries, bears, cats, swine, rodents, and various giant browsers and strong-jawed scavengers (the last two of lines that are extinct).

John Day Fossil Beds at Painted Hills

Evidence of fossil seeds, nuts, fruit, leaves and woody structures has also been found. Fingerprints of these are found in tuff formed from layers of volcanic ash. Other fossils are found in sandstones and siltstones.

The importance of fossils found here was first recognized in 1861 by Thomas Condon, who later became a professor of paleontology at the University of Oregon. By the end of the nineteenth century, scientists at the Smithsonian Institution were seeking fossils from this site, and soon those at other eastern universities. A few years later, John C. Merriam worked out the chronology of fossils found here. He was the first one who helped gain protection for these sites.

Subsequently a number of figures prominent in the evolution of the state park system spoke out to attract support for preserving the area, Sam Boardman and Robert W. Sawyer among them.

Once the National Park Service had affirmed the importance of this area, the state parks department worked out an agreement to transfer the sites from the state to the National Park Service. The state insisted on special provisions to protect its interests.

In 1975, by executive action the John Day Fossil Beds became a national monument under the National Park Service. Fourteen thousand acres are protected at three sites: the Painted Hills site northwest of Mitchell; the

Sheep Rock site northwest of Dayville; and the Clarno site west of Fossil. A major museum and research facility has been erected in the Sheep Rock unit, which is called the Thomas Condon Paleontology Center. A smaller museum nearby, known as the Cant Ranch Museum, tells the story of human occupancy in the area.

A first-class job is now done of interpreting the fossils of the area, which are of world renown in the field of paleontology. The Park Service is keeping faith with those who pushed for it to take over the sites. And those who worked out this transfer accomplished something of enduring importance.

Columbia Gorge National Scenic Area

John Yeon was not through with his efforts on behalf of the Columbia Gorge (see also Chapter 3). He lived on the Washington side and feared the new I-205 bridge would unleash more sprawl around him. In 1980, Skamania County had just approved a new subdivision across from Multnomah Falls and another just west of Beacon Rock. It was time to inject new energy into the efforts to protect the Gorge.

He knew Nancy Russell of Portland and her love of the Gorge. She had been spending a lot of time there leading groups and searching for wildflowers. He told her that it was time for someone younger to assume leadership of the effort to save the Gorge. He challenged her to weld the friends of the Gorge into a real organization that would press for its protection. She accepted his challenge and formed the Friends of the Columbia Gorge, with the encouragement of her hiking friend and wildflower specialist Russ Jolley.

They quickly decided that they needed new federal protection. But what kind? It did not fit the standard formulas. Clearly it had the kind of scenery and natural values of a national park, but it also had a lot of development—far more than most American national parks.

Could it be more like Canadian national parks that had towns in them? Chuck Williams, then of Friends of the Earth, thought so and actually got the U.S. National Park Service to do a study of the values found in the Gorge.

But most thought that was too much of a reach. They knew that the Forest Service had invented a category it occasionally used to give an added measure of status to areas that were both scenic and had roads and resorts. They were called "Scenic Areas." They could use that designation and work out a management prescription that Congress might bless. The approach would integrate the governments of both states and the local Indian tribes into the management of the new area.

For six years Nancy Russell led a campaign to get this designation. She spoke to group after group, raised funds, and lobbied politicians. She led the way through nine hearings, federal and state.

Her efforts often drew a hostile reaction from local residents; she faced denunciations, threats, and slashed tires. Some of them formed an opposing group: Columbia Gorge United, which particularly included property owners there. They were vigorous and outspoken.

Nonetheless support grew. The point came at which the Republican senators from the two states had to get involved in trying to negotiate a solution. Senator Slade Gorton of Washington state played a pivotal role as he persuaded many conservatives in Congress to accede. He played a price in the next election when many local conservative voters retaliated by supporting his Democratic opponent, Brock Adams, who replaced him (though there were many reasons for Gorton's defeat).

The various Democratic congressmen on the two sides of the river also had their own ideas of how to proceed, producing their own bills.

Finally they came to agreement on a bill, which they enacted and became law on November 17, 1986. Both Oregon senators, Hatfield and Packwood, pressed President Reagan to sign the measure, which he finally did at the last moment.

The key to achieving this agreement was inclusion of a provision giving equal status to economic development and natural values. Certain zones were designated for economic development, with a twelve-member commission composed of local residents put in charge of it. But no development was permitted that would compromise the natural, scenic, cultural, or recreational values of the scenic area.

Chapter 7—The Advent of New Reserves

A bi-state commission was given the job of designing the boundaries of the various zones in the scenic area. It was also to inventory the resources in the area and devise a management plan. It was to be composed of three members to be appointed by each state's governor and one to represent each county in the area: twelve in all.

In addition to the zone for economic development (at 30,000 acres), there was a zone where natural values were paramount—a Special Management Area of 115,000 acres, and a general management zone of 149,000 acres, where farming and logging could occur. The U.S. Forest Service was put in charge of administering all federal lands found there. The counties were put in charge of developing land use ordinances for private land. However, they had to be consistent with the management plan. As a result, no more subdivisions have been built since the bill was enacted.

The four tribes that had traditionally inhabited the area also had to be consulted about developments: the Yakama, the Warm Springs, the Umatilla, and the Nez Perce.

The final result was a statutory National Scenic Area of 292,000 acres running eighty-five miles along the Columbia from the mouth of the Deschutes River to Troutdale. The Scenic Area is unique in the way in which it blends the conflicting concerns and integrates the agencies and the Indian tribes into its management.

In the years following enactment, Nancy Russell and her organization bird-dogged all the implementing steps. She also focused on the need to acquire more of the inholdings. She was instrumental in raising funds to buy 40,000 more acres, and personally purchased thirty properties there that were in danger of development. The federal government has now spent $90 million on the Scenic Area, and stimulated the development of two tribal museums there. A spectacular new lodge and conference center has been built at Skamania.

The result is a trail-blazing new formula for protecting this special place that straddles the Columbia. Despite being born in controversy, Oregonians now see it as part of their heritage. Nancy Russell and John Yeon deserve to be remembered for their crucial role. They made it possible.

Newberry National Volcanic Monument

Few realize that Oregon has more than one Crater Lake. Two of them, Paulina and East Lakes, are in Newberry Crater south of Bend. As Oregon's largest ice-age volcano, Newberry Crater is five miles across. This volcano last erupted in AD 640. It was named after Dr. John S. Newberry, the geologist who visited this area as part of the 1855 Pacific Railway Survey.

Those who have known the area have long hoped it would get greater recognition and protection. While it was administered as a part of the Deschutes National Forest, the National Park Service did recognize its significance by placing it on the list of National Natural Landmarks in 1976. The superintendent of Crater Lake National Park even felt it should have been made a national park, administered as a separate unit of his park.

But boosters in Bend wanted to improve its status in a way that would engender less controversy. When Mt. St. Helens was made a national monument in 1982, Congress decided to keep it under the Forest Service. The Bend boosters then realized they could seek a solution along the same lines. They would aim at creating a national monument there under Forest Service management, which would be less controversial.

Since they would need the support of the congressman from that area, Republican Robert Smith, as well as Senator Hatfield, they sought to bring aboard as many community members as possible. They were led by Dr. Stuart Garrett, who had hiked widely in the area and knew the flora and fauna intimately. They did this through forming a committee of thirty members representing a diverse group of stakeholders in the community. These included not only conservationists and recreational users, but business leaders and local officials as well. The business leaders especially included geothermal power developers, since they posed the potential conflicting interests. Many of them were intent on drilling in the area to tap the scalding water that still lurked there around the underground magma.

The group spent thousands of hours over a three-year period trying to find a solution that everyone could accept. Representative Smith later characterized the process as one that did not exclude anyone, and one in which

Chapter 7—The Advent of New Reserves

Newberry Crater from Paulina Peak

no interest was viewed as superseding another. Local Forest Service officials provided support to the group.

Dr. Garrett saw himself as running a consensus process in which compromises were sought that would still allow everyone to support the final plan. The key to the final breakthrough was offering the main geothermal concern, the California Energy Company, alternative drilling sites nearby—but outside the boundaries of the proposed monument. A Special Management Area of 10,300 acres was set aside next to the monument for this drilling. If the sites were not developed, then in due course this acreage would be added to the monument without any need to go back to Congress.

Smith's staff refined their proposal further, which Smith then introduced as a bill. While the Democrats on the subcommittee handling the bill were generally complimentary, they did tighten its management provisions and shaped it so that it conformed to the standard format for such legislation.

One of them, however, Representative Rahall, expressed annoyance at all of the concessions made to the geothermal companies. He charged that the exchange provisions went too far and almost overwhelmed the bill. That explains why the Democrats, including Oregon's Representative AuCoin, did tighten it.

When it passed the House and moved to the Senate, Senator Hatfield took charge and steered it through. Being an admirer of consensus processes, he complimented Dr. Garrett, saying the process smoothed the way by negotiating solutions to problems. Representative Smith proclaimed that Dr. Garrett was "the father" of the monument. He has now written a book describing his experiences.

In 1990, a monument of 56,000 acres was established, which could be expanded administratively by another 10,000 acres.

While this monument is focused on its geological features, it has much of the diversity that typifies monuments. It is high enough to be clad in old growth forests, despite being in the desert; altitude offsets desert conditions. It has a variety of appealing features: clear deep lakes and an impressive waterfall; as well as abundant wildlife: pine marten, bear, and varied birds—eagles, osprey, and Tundra Swan. Rare plant species are found there too, such as the Pumice Grape Fern. And the Big Obsidian Flow is also an attraction; Indians fashioned arrow tips around its edges.

But some critics complained that Smith's bill made too many concessions to commercial interests: not only geothermal power firms but timber men and ranchers as well.

Twenty years later, controversies have erupted over whether geothermal drilling in the adjacent Special Management Area was something that was to be allowed on demand, or whether it was still governed by environmental laws of general application. The local unit of the Sierra Club has objected to drilling within two miles of the monument's boundaries. It fears that such drilling operations would be visible from high points in the monument, as well as warning about the risk of toxic effluents.

While one can debate whether too many concessions were made in the end, this is probably the only way the state could have gotten this fine monument at the time, given the outlook of its legislators. It is a valuable addition to the state's environmental assets. Dr. Garrett had the skill and knowledge to make this process a success.

CHAPTER 7—The Advent of New Reserves

Steens Mountain at Kiger Gorge

STEENS MOUNTAIN COMPLEX

Steens Mountain rises to almost 10,000 feet—a mile above the Alvord Desert. Creeks and rivers with fish in them, such as that in Kiger Gorge, flow to the north and east. Aspen trees cluster along its creeks flowing over its gentle, western slope.

This area, about seventy-five miles south of Burns, is geologically unique in Oregon. Cirques, carved in half circles, fringe its eastern escarpment. Some have called this glacially carved terrain "biscuit-board" topography. The Alvord Desert at its feet to the east and southeast is the product of the rain shadow created by the mountain. This is a dramatic example of fault-block topography—one of the best examples in the Great Basin.

And this is an area of exceptional botanical diversity. At various elevations, it harbors endemics, such as the Steens Thistle, and rare plants such as the Steens Paintbrush and Biddle's Lupine. It is located at the intersection of a number of eco-regions: it is at the western edge of the Rocky Mountain region; the northern edge of the range of species from California; and at the eastern edge of species from the West.

Biologists have always been anxious to secure protection for the area. On behalf of the Sierra Club, I pressed for better protection as a member of a local BLM advisory committee in the early 1960s. Senator Hatfield sought to have the area studied for its suitability as a national park in the late 1960s, with the Natural Desert Association supporting that designation. Hatfield later called for making it a National Conservation Area.

As the Clinton Administration faced its conclusion, it made it clear that it wanted to set aside a spate of new national monuments by executive order. At that time, Interior Secretary Bruce Babbitt visited the Steens and gave the local residents an ultimatum. Either Oregonians could work out a plan for the area that suited them—both ranchers and conservationists—or the Clinton Administration would do it. And that plan might not suit them. Coming from a ranching family in Arizona, Babbitt said he would prefer that they came up with a plan that kept them in the area. But standing pat would not keep the Clinton folk from working their will.

Babbitt gave the established BLM advisory committee for the area two months come up with a plan. But its mainly ranching members (fifteen of them) did not want to embrace a plan that curtailed their activities. Environmentalists were represented by the ad hoc Steens-Alvord Coalition.

So at the end of that time, negotiations were turned over to a smaller committee, formed mainly of staff from members of the state's congressional delegation, particularly Representative Greg Walden from eastern Oregon, as well as the governor. Environmentalists were represented by Andy Kerr, representing the Wilderness Society at the time, and Jill Workman of the state's Sierra Club chapter. Kerr took a tougher position, while Workman took a less intransigent position on grazing, singling out some ranchers who she felt were more responsible. But some environmental groups would not participate, being deeply skeptical that any progress could be made through negotiations, particularly the Oregon Natural Resources Council (ONRC) and the Natural Desert Association.

The key sticking points were the extent of cattle grazing and the use of off-road vehicles (ORVs). Eventually, the needed plan was hammered out, with give-and-take all around. The breakthrough was made in August of 1999, following public hearings at the county fairgrounds in Redmond earlier that month. With pressure from Congressman Walden, local opposition finally faded; Senators Gordon Smith and Wyden also played important roles. Provisions were made for various land exchanges and buy-outs of ranchers.

Once the plan was worked out, it went to Congress for approval, rather than being done as a national monument under the Antiquities Act. In the year 2000, the plan became law, establishing a series of breakthroughs.

This new unit was known as the Steens Mountain Cooperative Management and Protection Area of 425,000 acres. Of that, 100,000 were to be free of cows and cattle grazing, which was a first for BLM lands, and 175,000 acres set aside as wilderness. A federal wildlife reserve for Redband Trout was set aside in the Blitzen watershed on the west slope, another first. Wild and Scenic Rivers status was given to twenty-nine miles of streams there. And 1.2 million acres, including the entire Alvord Desert, were withdrawn from mineral entry, closing them to mining.

Following enactment, the BLM was slow in enforcing the new restrictions on grazing, ORVs, and others to protect its new wilderness there, but gradually the new reserves have taken form. They are among the most significant new reserves in Oregon. They deserve to be well managed and cherished.

The Steens complex did not come about because of an admirable consensus process. It came about because of a number of exceptional factors. The Clinton administration put them "under the gun." Then their congressman put the ranchers under pressure to come to terms or lose everything. And while some environmentalists would not deal, others would, and those that did had finely honed skills.

Once again, it demonstrates the power and inventiveness of Oregon conservationists.

CASCADE-SISKIYOU NATIONAL MONUMENT

Unlike the Steens, the Cascade-Siskiyou National Monument is not noted for its scenery (though it does have places of beauty). But it is important owing to its biological values. Both values are important in that regard, but this is the first national monument set aside under the Antiquities Act because it is an "ecological wonder."

Its 54,000 acres of land were set aside for this purpose under the BLM by President Clinton in 2000. Most of it is five miles east of I-5 along Oregon's southern border. While a third of it is still privately owned, these lands are being acquired by exchanges. It is anchored by Soda Rock (6000 ft.) and Pilot Rock (5900 ft.). In 2009, 24,100 acres of wilderness was set aside around these highpoints.

The area is pleasant, though, having old growth forests as well as lush wildflower meadows. Its streams are filled with trout and other fish found nowhere else.

This east-west mountain range just north of the California border is at a crossroads where species from the east, west, north and south of here meet. Ecotones such as this tend to harbor plants found only there (endemics) and this area does—Greene's Mariposa Lily and Gentner's Fritillary. Diversity is fostered by the variety of habitat niches found in the jumbled landscape of this area. Also it had not been heavily glaciated historically, allowing older, endemic species to survive.

This range connects the Cascades to the east with the Siskiyous to the west. It brings together species from the drier Great Basin together with those of the wet, deep forests of the coast and the Cascades. It serves as a gateway to one of the great reservoirs of biodiversity of the country in the Siskiyous. It was set aside, rather than the Siskiyous, because this was BLM land (under Babbitt at the time), rather than Forest Service land.

The monument is rich in birds, with over 200 species of them, including some that are endangered, such as the Great Grey Owl, the Northern Spotted Owl, and the Peregrine Falcon. Many are at the edge of their normal ranges; it is at the northern end of the range for the Blue-Gray Gnatcatcher; it is at the western limit for the Canyon Wren; at the eastern limit of the Hermit Warbler; and at the southern end of the range for the Ruffed Grouse.

It boasts of an almost unrivaled assortment of butterflies, usually found in very different ecosystems, because of the variety of host food plants.

While the impetus for this reserve came from Babbitt and the BLM, environmentalists did support its protection because it was being adversely impacted by cattle grazing and off-road vehicles (ORVs). The World Wildlife Fund did open an office nearby to build the case for it. Babbitt came to the area, as well as to the Siskiyous, to evaluate the sites and meet with concerned people. Strangely enough this did not then arouse much opposition. He chose boundaries that avoided heavy concentrations of valuable timber and mineral deposits—to minimize opposition. Incompatible grazing permits were to be retired, and grazing rights could be donated. A management plan for the monument was adopted in 2008.

After the fact, critics became incensed over the management plan and plans for reserving wilderness. They claim that the danger of fire is growing owing to lack of fire suppression; that roads are not being maintained; that timber companies have been pressured to bow out; that permit-holders are being pressed to relinquish their rights; and that owners of inholdings are being asked to sell. Those selling, though, do not seem to have objected.

Many interpret these complaints as really being aimed at blocking proposals to expand this monument along the Siskiyou Crest into California. Ideas have been broached of using executive powers to expand this monument by 200,000 acres. This expanded area would serve many of the same ends. We have not heard the end of efforts to extend greater protection to these "hot spots" for biodiversity in southern Oregon.

It remains to be seen whether establishment of this reserve marks a new trend toward protecting areas of primary significance for science, with the initiative coming from agencies and scientists, rather than from activists. At any rate, it opens an exciting, new and broader agenda.

Zumwalt Prairie Natural Area

Non-governmental groups are also beginning to set aside reserves of significant size. At Zumwalt Prairie on the west slope of Hells Canyon, The Nature Conservancy established a new reserve of 33,000 acres in 2000; 18,000 of

Zumwalt Prairie

these are rolling prairies. This is now the largest private nature reserve in Oregon, and a portion has been designated as a National Natural Landmark.

This is one of the largest unplowed bunchgrass savannahs on the continent. It escaped being plowed because of its remote location—between the Wallowas and Hells Canyon, with its short growing season at elevations between 3500 and 5500 feet. Among the native grasses that predominate there are Bluebunch Wheat Grass, Great Basin Wild Rye, and Idaho Fescue.

During the time that the area was part of a ranch, it was grazed but may not have been harmed. While the new managers are removing old stock ponds and check dams put in by ranchers, they believe that a certain amount of disturbance may be beneficial in recycling nutrients. OSU researchers are trying to determine the appropriate levels of grazing to maintain the ecosystem and to limit harm to biota and soils; some of them believe that low-intensity grazing is indicated, with the grazing animals rotated regularly.

But the intact grasslands are marvelously productive, supporting over 430 species of plants and forty-eight varieties of butterflies, as well as various ground-nesting birds. Cougars, black bears, and now wolves are also

CHAPTER 7—The Advent of New Reserves

seen here. Even one plant on the federal endangered list is found here: the Spaulding Catchfly.

The numbers of Rocky Mountain Elk here have exploded in a decade— from 500 to over 3000. Because use at these levels is damaging the range, managers are trying to spread their use over a wider area. It is still not entirely clear why the elk are concentrating on the prairie.

Reserves such as this reflect scientific impulses entirely. Activists are limited to the role of helping to raise the money to buy such lands. The Nature Conservancy now has over 83,000 acres of such reserves in Oregon.

Marine Reserves

As this was being written, Oregon began to establish its own Marine Reserves off its coast. Five have been established: one at Otter Rock north of Newport; one at Redfish Rocks near Port Orford; one at Cape Falcon (near Nehalem); one at Cascade Head (north of Lincoln City; and one near Cape Perpetua (south of Yachats). In addition to "no take" areas where no fishing can take place, they embrace buffers where no trawling is permitted. Altogether, they embrace 75,000 acres, or less than ten percent of Oregon's territorial waters.

They are designed to serve as nurseries for new sea life in the near shore zone. Industrial activity, and fishing and crabbing are barred in these reserves; drag fishing is particularly destructive. These sites were suggested by Oregon's Ocean Policy Advisory Council and were authorized by the legislature in 2011 and 2012.

The last three were set aside in 2012 in the hope that the Our Ocean Coalition would not put a broader initiative on the ballot as a response to inaction (polls showed three-fourths of the public favored such reserves). The Oregon Conservation Network took the lead in lobbying to get the legislature to do this. Fishing groups had divided views on the idea of marine reserves. Governors Kitzhaber and Kulongoski strongly supported establishing them. In 2022 scientists will review the effectiveness and impact of these reserves.

Lesser Reserves

Two reserves were set aside in Oregon in the early 1970s under various established programs.

One was the Cascade Head Scenic Research Area of 9670 acres on the Oregon coast, which is a unit of the Siuslaw National Forest. It was given this unique designation by Congress in 1974. Normally, either Scenic Areas are established administratively, or as Research Natural Areas. Half of it had begun in 1934 as the Cascade Head Experimental Forest, which includes another 6000 acres. This is something of a hybrid that originated with the Oregon delegation. Over 350 species of wildlife are found there, including the rare Cascade Head Catchfly and the Oregon Silverspot Butterfly (which depends on the Early Blue Violet). It is also designated as a United Nations Biosphere Reserve.

Soon after Congress established the system of National Estuarine Areas (in 1972), a unit of it was established administratively in Oregon: the South Slough National Estuarine Research Reserve of 4800 acres, south of Coos Bay. This is one of the most accessible of the units in the system and features an inviting visitor center on its west side. It is jointly managed by the Division of State Lands and the National Oceanic and Atmospheric Administration (NOAA).

Both are viewed as reserves of lesser importance. This might change if unexpected values are discovered there, as has been the case with the Oregon Caves monument.

In addition it is worth noting that the BLM has established over 200 Areas of Critical Environmental Concern (ACECs) in Oregon, of which some are Research Natural Areas and some are known as Outstanding Natural Areas. Some are for outstanding geological features (such in the Diamond Crater area of eastern Oregon— known for its diverse basalt formations); some are in the Coast Range exemplifying various old growth trees (e.g., Neskowin Crest) and protecting habitat (Elk Creek safeguarding nests for Bald Eagles); and some protect habitat for endemic plants (as in the Rough and Ready Flat area near Cave Junction, which is on serpentine soils). While the ACECs are set aside under statutory authority, they are administrative creations, but they do seem to have a degree of stability. While most are relatively small, they do add up to over 172,000 acres. The Forest Services' various special areas include over 130,000 acres, of which 78,000 acres are in Research Natural Areas (ranging from 100 to 1000 acres).

Reassessing the Oregon Caves National Monument

For many years, conservationists thought of Oregon Caves National Monument as a reserve of only limited importance. It was small—only 488 acres; and its caves were limited in extent. In effect, it was regarded as a minor reserve.

Now new discoveries there have changed how it is seen. While its caves are not large, they are among the finest of the caves of marble, as contrasted to limestone. And important fossils have been found in its deepest chambers, particularly Pleistocene jaguars.

But it is in the biological realm that it now shines. An impressive array of insects live in its caves—in fact, one of the largest collections anywhere of endemic insects living only in caves.

And outside, on this small tract of forest land over 400 species of plants have been found. Among them is a Douglas Fir with the greatest girth in Oregon—at thirteen feet— and an estimated age of 1500 years (at 4800 feet elevation).

Important steps have also been taken to restore the habitat for the cave-dwelling insects. The natural flow of air has been restored by installing airlocks, and the streams flowing through the caves now run clear again with the accumulated rubble removed (the product of visitors and past management notions). The build-up of algae is now being controlled with bleach, with insects from the surface being kept out of the caves.

These caves were originally set aside as a national monument by President Taft in 1909, following publicity by Joaquin Miller, who visited them in 1907. Following discovery in 1874, they became a tourist attraction.

Articles in the *San Francisco Examiner* in the early 1890s dubbed them as the Oregon Caves to distinguish them from caves in northern California.

First managed by the Forest Service, they were transferred to the National Park Service in 1933. The CCC built the widely admired chateau there in 1934.

This is not the first time attitudes have changed as holdings have been better explored and understood. But this is a first for a reserve in Oregon.

References

The material in this chapter was derived from various online sources and interviews with people who were involved.

CHAPTER 8
PROTECTING WILDLIFE: REFUGES AND PROGRAMS

WILLAMETTE VALLEY FEDERAL WATERFOWL REFUGES

Since the mid-1960s, a string of national wildlife refuges has sprung up along the Willamette Valley. Four of them now dot the valley from Corvallis north—and more may be coming. They began with the Finley refuge in 1964, with Baskett Slough and Ankeny both coming along the following year. And Tualatin River came into being in 1992, and it is still being fleshed out (see discussion below). Altogether, they now protect 12,262 acres of habitat.

All of them are designed to provide wintering ground for the dusky subspecies of the Canada Goose. These geese fly north in the spring to spend the summer on the flats of the Copper River Delta in Alaska. Once they have flown north, visitors are welcome. Some of the refuges feature natural wetlands, while in others they have been engineered.

But in places, some of these refuges protect other types of habitat, which are becoming rare. Some of them (Finley) have disappearing Valley Oak Savannah, where endangered plants are found such as Golden Paintbrush and Kincaid's Lupine. A butterfly once thought to be extinct has been found on Baskett Butte—Fender's Blue. The endemic Peacock Larkspur is found on the Finley refuge. It is among the lush wildflowers found on the 400 acres of wet prairie there that have never been plowed because of their high clay

Finley National Wildlife Refuge office

content. The Baskett Slough has one of Oregon's largest concentrations of the Streaked Horned Lark, which is on the list of species in critical condition in the state.

Because these refuges undertake these multiple jobs they are regarded by the Fish and Wildlife Service as the "crown jewels of the Willamette Valley." They protect some of the best and rarest habitat in the valley, as well as supporting waterfowl.

How did these refuges come about? It is not simply enough to say that the Fish and Wildlife Service bought them with Duck Stamp and hunter tax money. Someone provided the drive and insight to bring them into existence, and public support was needed.

There was such a person. He was David Marshall. And he got support at a critical point from the Portland Audubon Society and the Oregon Duck Hunters Association.

Marshall was an Oregon native who grew up in the Portland Audubon Society and in the tradition of William Finley. In retirement, he wrote the definitive guide to Oregon's birds, as well as Oregon's non-game wildlife management plan. But he was also a career biologist with the Fish and

Wildlife Service. In 1960, the national office of the service put out a call for suggestions for qualified additions to the national refuge system. Marshall was then a biologist at the Malheur refuge who just happened to know some ideal areas.

At an earlier time, while he was earning his degree at Oregon State, he had done a study on the case for a refuge for the Dusky Canada Goose around Muddy Creek south of Corvallis. He had been looking at a 3500-acre tract north of McFadden's Marsh, which was also known as the Failing Estate. At that time, he had even contacted the owner and had been encouraged to proceed with his studies on the estate. He had also become aware of the Baskett Slough area north of Rickreall, which supported a large population of geese.

Marshall put forth suggestions for both areas to the national office. They checked out his ideas, found them to be sound, and then encouraged him to proceed.

But obstacles loomed. When he then approached the owner, Henry Failing Cabell, Cabell made it clear that he was enjoying use of the property and had no interest in selling. However, a few months later, Cabell had reconsidered. Having no heirs, he wished to see the tract preserved and would sell it for a refuge, but he did not want to break it up by selling just the wetlands.

Marshall faced the problem that his funds were supposed to be used for acquiring wetlands. Could he buy non-wetlands as well when they were part of the parcel that could only be bought as an entire package?

At that point, he needed to put together a comprehensive proposal for the valley that would permit their Washington office to ascertain the need. He would need to justify the acquisitions and use of the funds, and propose specific boundaries. Marshall put together a package that included the entire Failing Estate, McFadden's Marsh, Pigeon Butte, and the Baskett Slough tract (including some Oregon White Oaks on the north). All together, it embraced 7000 acres.

While his proposal was audacious, it was subsequently approved by his superiors in Washington, D.C.

But he had trouble getting approval at another level. At that time, authorities in each state had to approve plans to establish new federal refuges there. In this case, in 1962, the director of the state Game Commission told him that they wished to limit the impact on the tax rolls of each county. In the case of the Muddy Creek proposal in Benton County, they would not approve federal acquisition of more than 5000 acres. And elsewhere throughout the valley, acquisitions could not exceed 2500 in any county. The Game Commission also proposed spreading the acquisitions over five units in the valley.

Opponents also began to be vocal, with strong criticism coming from a local agricultural extension agent and a major lumberman. But supporters also emerged: one was a wildlife extension agent, and others came from the Oregon Duck Hunters Association and the Oregon Audubon Society (now the Portland Audubon Society). A public hearing in January of 1963 in Corvallis proved to be particularly stormy.

At that point, the opponents went to their legislators and persuaded them to push forward a bill that required each county to grant approval before the federal government could acquire land for a new refuge. And the legislature actually passed this bill, which threatened to strangle the whole idea in red tape.

To keep this from happening, the Portland Audubon Society and the Duck Hunters Association asked to meet with the governor, who was then Mark Hatfield. They got the meeting and told the governor why they thought this idea was bad public policy and why he should veto the bill. He agreed and did veto it. The way was cleared.

Marshall actually identified five suitable areas, but was only able to move forward on three of them at that time: the Finley unit, the Baskett Slough unit, and the Ankeny unit. While the latter one did not then support geese, water could be brought to it for geese, and it was inexpensive.

The legal basis for acquiring uplands around wetlands was subsequently scrutinized by the General Accounting Office (GAO, the federal auditing agency). Marshall stated his case—i.e., that they were all tied together in the parcels for sale. He was vindicated, as well as supported by his superiors.

So the emergence of these refuges was anything but a routine bureaucratic operation. Dave Marshall was the father of this splendid system. And citizen support emerged at the right time to rescue it.

Disputes over Grazing in Malheur Refuge

Given that Marshall's time at Malheur has already been mentioned, it should be noted that disputes subsequently erupted over the degree of grazing there at the time. But this was attributable to decisions made by its first and long-time manager, John Scharff.

He made decisions that cattle grazing could serve as a management tool and allowed cattle numbers to grow from 21,000 animal units per month (AUMs) in 1941 to 125,000 AUMs in 1970. That level of impact led to vocal complaints by visitors in the 1970s, which included Denzel Ferguson, who became a noted critic of grazing on public lands.

They noted a severely trampled landscape that looked like the worst of the days of the cattle kings at the refuge. The outcry of the public produced changes, resulting in the level of grazing today being one-fifth of that in 1970. It is restricted to meadows in the fall, and is only used when burning is not feasible. Moreover, cattle are kept out of streams.

However, they are still dealing with all of the barbed wire that cattlemen strung in the refuge. Hundreds of miles of it are still there and harm wildlife (flying birds and pronghorn) and are slowly being removed. Groups such as the Corvallis Audubon Society are volunteering their time to laboriously pull it out.

For the most part, I do not deal with changeable management decisions in this work. However, this case dealt with a matter so egregious that it is well remembered by long-time conservationists. Because of this controversy, the decision to reduce abusive amounts of grazing is not likely to be reversed.

Tualatin River National Wildlife Refuge

This refuge just south of Portland, between Tigard and Sherwood, is unique in being an urban refuge. While it was contemplated originally by David Marshall, little progress was shown for decades. The reason was that there was internal resistance in the agency because it was in the path of urban development and showed signs of urbanization (e.g., power lines and highways). It was not the customary setting for refuges.

But in the early 1990s, the national office embraced a new policy that encouraged development of refuges near cities, where public outreach and education could easily occur. This was just such a new kind of refuge.

And the citizens of Sherwood showed that they wanted a refuge along the bottoms of the Tualatin River. In 1991, their city council passed a resolution asking Congress to bring this refuge into existence, and they got other nearby communities to support their quest: Tualatin, Tigard, and King City. Crucial support was provided by a group of activists known as Friends of the Refuge, headed by Joan Patterson.

Senator Mark Hatfield welcomed their petition and secured approval of the idea in 1992. He also later secured funding to build a handsome visitor center for public education just off highway 99W. Representative David Wu helped.

Slowly lands that have been farmed along the 100-year floodplain of the Tualatin River have been acquired, with some even being donated (by Tom Stibolt). Three hundred forty acres have now been restored. Ultimately, over 4000 acres will be acquired; about forty percent of them have now been obtained.

By 2006 enough land had been acquired to open the refuge, with the visitor center built a few years later. Ralph Webber was the first refuge manager, who began the painstaking work to put this refuge together, with just his pickup and lots of determination.

Now over 50,000 ducks and geese and 200 species of birds are seen here. It is an important breeding ground for songbirds migrating to the tropics and for wood ducks. And the dream of the citizens of Sherwood is still being realized.

STATE WILDLIFE MANAGEMENT AREAS

Since 1944, the state of Oregon has been using grants from the federal Pittman-Robertson Act to acquire habitat for what are generally known as Wildlife Management Areas. This Act imposes an excise tax on firearms and ammunition that hunters pay to acquire areas where they will be allowed to hunt and can find the wild animals they seek.

Oregon now has twenty such areas, totaling over 200,000 acres. Most of them are wetlands for waterfowl, but a few are upland areas for other wildlife. The first was Summer Lake in eastern Oregon (between Bend and Lakeview). Flocks of Snow Geese winter on the marshes there. Another well-known one is on Sauvie Island, just west of Portland. Hundreds of thousands of waterfowl use this area. Both have about 12,000 acres. They are open to the public during the non-hunting season (the summer months); parking fees are beginning to be imposed. They are managed by the state Department of Fish and Wildlife.

State Wildlife Management Areas serve a useful purpose, but I do not accord them as much weight as federal wildlife refuges. Why is this? Both systems serve similar purposes and are funded largely from the same sources: the federal Duck Stamp Act of 1934 and the 1937 Pittman-Robertson Act. These grants are subject to federal oversight.

However, there are some differences. The federal system of wildlife refuges antedates these acts and is under-girded by more law.[30] Moreover, the amount of acreage in the federal refuges that is open to hunting is limited; no more than forty percent of the acreage is open to hunting. It exists primarily to provide a home to wildlife.

In contrast, all of Oregon's Wildlife Management Areas (WMAs) are open to hunting. While many other areas are used by hunters (where land

---

30 The following laws do this: Migratory Bird Act (1913), Migratory Bird Hunting and Conservation Stamp Act (1934), Fish and Wildlife Act (1956), Duck Stamp Act (1958). Refuge Recreation Act (1962), National Wildlife Refuge System Administration Act (1966), National Wildlife Refuge System Improvement Act (1997). Also relevant is the Migratory Bird Treaty (1916), interpreted in *Missouri v. Holland*.

Summer Lake Marshes

owners permit hunting), Oregon's WMAs are available to them without question. They are also used for other purposes, such as bird watching and hiking.

In addition to being open to hunting, Oregon's areas are actively managed and manipulated to encourage use by waterfowl. They are less natural than the federal refuges.

And federal refuges are subject to the "compatibility" test; i.e., only those activities are allowed that are compatible with the habitat needs of the wildlife. This test makes the needs of wildlife the predominant purpose of federal wildlife refuges.

These differences tend to give the federal refuge system more status and stability.

But are states unlimited in what they can do with these lands? Can they "play fast and loose" with them? In other words, can they eliminate portions of a unit when they desire—e.g., when conflicts arise?

This question may not have been tested. In such a case, conservationists could argue that state systems of Wildlife Management Areas are subject to a Public Trust responsibility; i.e., that state authorities that have received these funds must continue to use the acquired lands for this purpose and no others as long as they are relied upon by wildlife. In other words, it could be argued that states have limited discretion to add and subtract from areas acquired for this purpose.

Oregon's system of Wildlife Management Areas seems to have been relatively stable and well accepted. But their status under the law is less clear.

STATE EFFORTS TO PROTECT ENDANGERED SPECIES

At the behest of the Native Plant Society of Oregon, the legislature in 1987 enacted the Oregon Endangered Species Act. It does not impose any obligations on owners of private property, nor on the federal government. It seems to only impose obligations to safeguard these species on lands under the control of the state and cities and counties, as well as on Metro (managing certain functions and open space in the three counties in the Portland metropolitan region).

Oregon maintains only a tiny staff to administer its system, with responsibility divided between two agencies. The Department of Fish and Wildlife handles animals, and the Agriculture Department handles plants. While Oregon has its own list of species, it also incorporates the federal list.

On its own list, Oregon lists seven animals as endangered, and thirteen as threatened. It lists thirty-one plants as endangered, and thirty as threatened. It also maintains a list of 100 species on a Watch List because it is believed that that they are in decline. The Division of State Lands also maintains a data base of rare, threatened, or endangered species in Oregon, as well as of ecosystems and vegetative types.

While the state is supposed to be developing recovery plans for species just on its own list, it is not clear that much is being accomplished. Arguments over the federal system have probably discouraged it from investing much effort in its system.

Portland's Heritage Tree Program

While a number of cities in Oregon have programs to celebrate their outstanding trees, Portland has a program that actually protects them. And it protects a lot of them. As of 2013, it had 290 trees in its program, with over half of them on private property; the rest are in parks or on rights-of-way.

Every such tree is designated specifically by the City Council, and each owner must care for the tree and not allow it to be harmed. In Portland's system, this designation is extended to trees that are regarded as important because of their age, size, type, horticultural value, or historic associations. Most are given this designation because they are old or large, but 120 species are represented in its system.

Before a privately owned tree can qualify for this status, the owner must consent, but a lot of them have—158 as of 2010. They are owners who are devoted to these trees and want to see them safeguarded as others succeed them in ownership. When they consent, caring for them becomes a condition attached to the title; this condition will show up on future title reports. The city forester is in charge of enforcing these obligations.

The program achieved its current form in 1993 under the leadership of Jane Glazer.

One of the oldest trees in its system is the Council Oak, on the west side of Council Crest. This Oregon White Oak is probably over 300 years old. It also holds the record for the oak of this type with the broadest spread of its crown.

Trees and their heritage have become a part of the culture of Portland. And this program provides a popular way to introduce people to the importance of trees, in all their variety, to their lives.

References

David B. Marshall, *Memoir of a Wildlife Biologist* (Portland Audubon Society, Portland, 2008).

Material in this chapter was also derived from interviews and online sources.

CHAPTER 9

BREAKTHROUGHS ON THE NATIONAL FORESTS

STRUGGLES OVER BULL RUN

Portland has long enjoyed water of pristine quality from the watershed of Bull Run, a watershed just west of Mt. Hood.[31] At the outset, its managers went to lengths to keep it from becoming contaminated. They did not believe it was appropriate to permit any logging there, and even kept hikers and other visitors out. This approach began with its first superintendent in 1897, and this policy persisted for over half a century.

Bull Run has always remained part of the national forests there, while being reserved as a water supply for the city of Portland. Management has always been divided between the two entities, in an awkward arrangement.

In the 1950s during the Eisenhower administration, the Forest Service fell under new management that put more emphasis on serving commercial goals. Among other changes, it decided that municipal watersheds would no longer be off-limits to logging. Claiming that the stands there were fire hazards because the stands were supposed to be decadent, it persuaded Portland's mayor and city council that commercial logging should occur there to remove them.

---

31 It was the first area of federal forest withdrawn for special purposes in Oregon (1892).

Michael McCloskey | 173

Thus, access roads were pushed along the slopes across about a third of the watershed to facilitate this logging, which began in 1958. Over the next fifteen years, some 16,000 acres were stripped of their forests—even while the public was excluded. Commercial tractors and trucks moved across the 300 miles of logging roads that came to crisscross the basin.

Influenced by the Forest Service, Portland's Water Bureau insisted then that the logging posed no threat to the quality of the water coming off these slopes. And much of the public came to believe that the trees were removed in a sensitive manner with horses, with the manure being caught by a kind of diaper. Newspaper publicity suggested this. But it was not the case—tractors were used.

In 1973, a retired Portland doctor who had a nearby cabin observed a procession of logging trucks streaming down from the area. Outraged, Dr. Joseph Miller formed a citizens group to investigate and protest—BRIG (Bull Run Interest Group). Initially, he and OEC sued the Forest Service in federal district court for violating the Bull Run Trespass Act of 1904 and won (in 1976). The court not only held that commercial logging violated that Act, but the judge even held that the forest was not being properly protected. For those reasons, he halted further logging in the basin.

In response, Senator Hatfield and Representative Robert Duncan, then representing Portland, had Congress intervene with new legislation—ostensibly to provide for proper management of the area. Representative Les AuCoin secured a provision that he hoped would protect water quality. But actually, the new legislation merely legalized the ongoing logging under the guise of correcting the situation. It was backed by the city council, which strongly defended the logging.

Then the situation got worse. A windstorm in 1983 flattened forests around the edges of the logged areas. The response of the Forest Service was to cut down even more of the forests there under the guise of salvaging the down timber. Another 6000 acres were cleared. A scientific advisory committee, which was set up to evaluate the compatibility of logging with water quality, was split and could not reach agreement.

Unusually heavy rains hit the watershed in 1996, producing muddy runoff from the clear-cuts and access roads. Mud was now in colloidal suspension in its Bull Run water. To avoid delivering muddy water to the homes of Portland, the city had to turn to its auxiliary supply source, the well-fields along the Columbia. The policy of allowing logging and roads in the watershed was not working to keep the water clear. The situation was getting worse.

So at last in desperation the mayor and city council asked Senator Hatfield to obtain legislation firmly and finally putting an end to the logging there. In 1996, this was enacted, ending twenty-five years of contention over this issue. And in 2001, the watershed management unit was enlarged as Congress added the drainage of the Little Sandy River; the entire management unit is now closed to logging.

The city Water Bureau now has had to contend with the legacy of all these logging roads. For a number of years, muddy waters would come off the area during peak storms. Painstakingly, the city has had to repair the damage done then. Most of it has now been done, but, even though many of the old roads have been paved, they still need to be repaired when they fail and slump in storms.

And it is now clear that the danger of summer fire from lightning strikes was over-stated. Because of the very moist summer conditions in the basin, burns seldom occur.

And it is now clear too that the forest has other values besides as a water source. Along portions of the stream without water storage dams, it provides habitat for rare and endangered plants that OSU researchers have found.

It needs even greater protection.

Growing Conflicts over Logging Roads

In the 1960s through the 1980s, logging roads and clear-cuts were breaking out all over Oregon's federal forests. Many conflicted with recreational values and compromised the environment. One stands out since it involved a much loved hot spring—Bagby Hot Springs in the upper Clackamas River drainage, where the cedar soaking tubs were visited after a walk through a stately old growth forest.

In 1962, Larry Williams (later OEC's founder) and friends encountered a logging road being built into that forest. On visiting the supervisor to protest, they were told that the road would go no closer to the springs. But on visiting the area a week later, they found that the road had been pushed in another half mile. Armed with documenting photos and maps, they visited the supervisor again, but again were told it would go no further. But again, they found it kept creeping closer. On the third visit, the supervisor finally revealed that the Forest Service planned to push the road to within a quarter of a mile of the springs, notwithstanding his earlier assurances.

It was only when he wrote a letter of protest to the *Oregonian*, which it published, that the supervisor finally phoned to admit that they had decided to change their plans and to re-route the road elsewhere.

In many ways, this story epitomizes the endless conflict over roads and logging and explains why environmentalists put so much emphasis on gaining wilderness classifications. It seemed to be the only broadly applicable designation that could keep logging destruction at bay.

Adding to the Wilderness System in Oregon

1978 Wilderness Additions, Including French Pete Creek

Oregon solved some of its intractable wilderness disputes in 1978 with the enactment of the Endangered American Wilderness Bill. It included a component put forth by Senator Hatfield called the Oregon Omnibus Wilderness Act and another championed by Senator Robert Packwood, which focused on the unit in the Willamette National Forest known as French Pete Creek. The latter embodied the surviving remnant of the 53,000 acres eliminated from the Three Sisters Wilderness Area in 1957. Representative Jim Weaver from southwestern Oregon also fought tenaciously for the French Pete Creek unit.

Hatfield's bill began in 1976 with eleven proposed wilderness units around the state, but had shrunk to four by 1977. At that time, his package included units at Boulder Creek in the Umpqua National Forest, additions to the Kalmiopsis Wilderness in the Siskiyou National Forest, additions to the Mt. Hood Wilderness at Zigzag Mountain, and the Wenaha-Tucannon Wilderness in the Umatilla National Forest. He made major reductions in his bill notwithstanding a three-to-one margin of support for his larger bill at field hearings in Grants Pass in October of 1976.

Statewide support for the omnibus bill came primarily from environmentalists, especially the Oregon Environmental Council and Larry Williams. Opposition came largely from timber operators, though the small operators association did support the idea of moving wilderness ideas through the vehicle of an omnibus bill.[32] At hearings in 1977, some new issues began to surface, particularly regarding additions to the Kalmiopsis Wilderness in southwestern Oregon. The environmental side emphasized the uniqueness of plants in the area and their importance to science, while the opponents emphasized the value of the area for its rare minerals and the opportunity to mine them.

Hatfield was seen then as wanting to find a middle ground between the timber industry and environmentalists. He thought he could do this by advancing some proposals for wilderness that the environmentalists wanted, but far from all of them. By putting together packages of proposed units, or omnibus bills, he felt he could best position himself in this middle ground. He would pick and choose the units that he felt were politically viable. Eventually, environmentalists concluded that Hatfield was emboldened to move bills for wilderness and wild rivers mainly when he faced elections.

At that time, Senator Packwood was experimenting with currying favor with environmentalists and was trying to distinguish himself from Hatfield. In the late 1960s and early 1970s, the future of the French Pete Creek area had become a cause célèbre in the Eugene campus community. It was seen as one of only three forested valleys of ten miles or more in Oregon that had

---

32 Support came from the executive officer of the Western Forest Industries Association.

CHAPTER 9—Breakthroughs on the National Forests

then not been logged. A number of marches were held to emphasize their call for saving it. Fifteen hundred or more marched in the first demonstration in 1969. I marched in the second in 1971. At one point, its defenders were reduced to holding a mock Congressional hearing in Eugene.

This campaign was unique in that it combined two tactics that were rarely joined—direct action protests with legislative strategies. Over the twenty-five years that the valley was in contention, almost every tactic had been tried by its advocates. For a while, the focus was on administrative steps; then it was on legal action; then it was on keeping hope alive; then it was on finding ways to buy time; then it was on protests; and finally, it was on legislative action.

Long-term leadership of the French Pete campaign came from the local Sierra Club unit in Eugene, led at first by Professor Richard Noyes and then later on by Holway Jones. During the early 1970s, student groups on the U of O campus provided the spark that produced the marches.

In the late 1960s, Senator Packwood actually introduced legislation to classify the area as an Intermediate Recreation Area, which would have steered management away from broad-scale logging but allowed removal of dead and down timber and provided some recreation facilities. This represented the first turn away from broad-scale logging there by a prominent political figure. Finally, in 1973 Packwood turned all the way to the wilderness solution.

Hatfield would not agree to Packwood's approach for an intermediate designation. But he did ask the Forest Service in 1972 to delay putting in logging roads to look at another alternative: using helicopters to log.

As the issue became embroiled in tensions between the senators and then partisan politics, the Forest Service came to realize that it would probably never be able to fully implement the decision made in 1957 to log the entire area. Some of them probably even welcomed this turn of events.

As Hatfield prepared to push his omnibus bill out of the Senate in 1977, the House of Representatives pursued an entirely different course. At the behest of the national wilderness groups, particularly the Sierra Club and its conservation director, Doug Scott, Representative Morris Udall (then

French Pete Vista from Lowder Mountain

French Pete Creek

CHAPTER 9—Breakthroughs on the National Forests

chairing the Interior Committee) had the House pursue an omnibus bill for the entire country. It focused on what then were known as de facto wilderness tracts that had not been given high scores in Forest Service evaluations, such as RARE II, and were thought to be in danger of soon being logged. This package of state-by-state bills was known as the Endangered American Wilderness Act of 1977 and embraced an amazing total of 1.3 million acres. In most affected states, political problems had been resolved and a consensus reached to set them aside.

Udall's bill included Oregon units that had not been in Hatfield's measure, including French Pete Creek, one called the Middle Santiam, and one known as the Wild Rogue, which was different than Hatfield's Kalmiopsis additions. Congressman Weaver particularly championed inclusion of French Pete Creek and the various additions along the Rogue and adjoining the Kalmiopsis.

Three hearings were held in Washington, D.C., in the spring of 1977 on Udall's bill. At some, panels of Oregonians spoke in support and opposition. Most covered well-trod ground, but there was more focus on the Kalmiopsis additions and on the Wenaha-Tucannon proposals. A letter of support for French Pete Creek was introduced for the record from Oregon Governor Robert Straub, and Congressman Weaver was a vigorous presence at them for French Pete Creek and the Kalmiopsis additions. In due course, it became clear that President Carter supported Udall's measure.

Hatfield secured Senate passage of his bill in June of 1977, and Udall's bill passed in September of that year. Udall's bill did not include the Boulder Creek unit, supposedly to provide space in a conference bill for French Pete Creek. Hatfield's bill did not include the Middle Santiam unit or the Wild Rogue.

Udall's committee report also pressed the Forest Service to relax its "purity" standard that disqualified roadless areas exposed at their borders to the sights and sounds of civilization. This standard had led to the Endangered Wilderness Act. Then Assistant Agriculture Secretary Rupert Cutler intervened to direct the Forest Service to change its approach.

Late in the fall of 1977, the House and the Senate struggled in a difficult conference committee to reconcile the bills. Congressman Weaver was the key House negotiator and won his aims: the French Pete Creek unit at 45,400 acres and the Wild Rogue Wilderness at 36,700 acres. Hatfield got three of his four units: the Kalmiopsis additions of 92,000 acres, the additions to the Mt. Hood wilderness of 33,000 acres, and the Wenaha-Tucannon wilderness of 180,000 acres (it straddles the border with Washington state).

The number of acres chosen by the conferees for the additions to the Kalmiopsis Wilderness was greater than that in Hatfield's latest bill, but far less than Weaver got put into the House bill. At 92,000 acres, it was a compromise that gave only a little ground to Weaver. Weaver was the first Oregon congressman to enthusiastically embrace the environmental agenda.

Hatfield had to defend his Mt. Hood additions against efforts to shrink them, as well as efforts by Representative Ullman to eliminate all timber-rich areas from the Wenaha-Tucannon Wilderness. In both cases, he prevailed.

Boulder Creek and the Middle Santiam units did not make it at this time. Timber operators in Douglas County launched a strong campaign against the Boulder Creek unit. Both of them became designated wilderness at later times.

A total of 387,000 acres of wilderness were established in Oregon in 1978 by this measure. It brought to an end the long controversy over the French Pete Creek issue and the 1957 decision to turn over the best forests in the Three Sisters Wilderness to logging. With that decision in mind, the difference was finally split between the contending forces.

Wilderness Designated in 1984

As it turned out, even more wilderness was designated a half dozen years later— in fact twice as much. In the Oregon Wilderness Act of 1984, 861,500 acres in Oregon were put in the national wilderness system in thirty-one units. Twenty-three were new units, and eight involved making additions to existing units. The largest were the North Fork of the John Day in eastern Oregon and the Sky Lakes Wilderness along the Cascade crest south of Crater

Lake National Park. The latter was similar in size to the one I had laid out in a study in the early 1960s. A new Cascades Recreation Area of 86,200 acres was also established in the Mt. Thielsen–Diamond Peak area.

These designations were a curious mixture of traditional, high country areas, beloved by backpackers, and a new kind of wilderness: old growth forests of interest to naturalists. Some of the naturalists came from the Lane County Audubon Society, and much of the support came from the relatively new Oregon Natural Resources Council (ONRC). Those who were focused on old growth viewed it as of critical importance to the welfare of rare and endangered fish and wildlife. But in a way, this shift in perspective began with the French Pete Creek campaign, which in no way focused on "wilderness on the rocks." The fact that the Sierra Club led this campaign suggests that the organization was already moving out of the high country.

While those with the perspective of backpackers, from groups such as the Sierra Club, and those with naturalists' interests, from groups such as ONRC, were both intensely involved in this campaign, they did not work well together. They marched to different drummers, and did not always agree on the strategy for how the campaign should be run. Part of this tension grew out of the fact that while ONRC was concerned just with one state (Oregon), the Sierra Club was part of a national organization concerned with overall national strategy on wilderness.

But, ONRC was producing a new source of energy and talent that wanted to displace the Sierra Club; they did not to turn to it for political advice. They were rivals for leadership. These frictions embodied conflict at a number of levels: a cultural conflict, generational conflict arising out of experience with such issues, and interpersonal conflict because of these. It was ironic that ONRC became so critical of the Sierra Club because it grew out of the Oregon Wilderness Coalition, founded by Sierra Club leader Holway Jones.

There were also very serious issues at stake. Two sources pointed to where the remaining wild areas were that had survived. These were the Forest Service's second roadless study (RARE II), which identified 2.9 million acres of roadless tracts in Oregon's national forests, and a research paper by Cameron La

Follette which inventoried all existing old-growth from the Forest Service's own records. Prepared as a senior's thesis at Reed College, it was entitled "Saving All the Pieces." With its findings in mind, she and Sydney Herbert, on behalf of the Lane County Audubon Society, then triggered a series of evening seminars on old growth by experts drawn from universities and agencies, culminating in a well-attended conference to educate the public, which agencies also sponsored.

Both organizations were outraged that the Forest Service had only recommended saving 415,000 acres of these for protection as wilderness; most of the rest would be made available to be roaded and logged. They believed that the RARE II process was deeply flawed and felt that its inadequacies should be challenged in court.

The national Sierra Club had helped identify RARE II's inadequacies, but had grave reservations about going into court to challenge it. While recognizing that one might score an easy court victory because of its poor EIS, they and the Wilderness Society also feared that such a victory might trigger moves in Congress to quickly override it and, if enacted, make them worse off. This had happened a few years before in West Virginia in the Monongahela timber marking case, and they did not want it to happen in Oregon or affect any of the bills likely to come along in other states. The Club's conservation leader, Doug Scott, was very vocal about this peril, which was resented by ONRC's leader, Andy Kerr, who was emerging as a canny tactician.

Local Sierra Club leaders were caught in a bind. While some believed that the lawsuit could push things along, others agreed with the Club's national strategy of not suing and thereby risking a backlash in Congress, triggered by industry. These chapter leaders were also mindful that countless wilderness proposals in other states might thereby be jeopardized. Wilderness in Oregon was not the only acreage at stake. And this tactical debate was taking place in a political context that had changed: Ronald Reagan was now president and his appointees were pushing heavy amounts of national forest

CHAPTER 9—Breakthroughs on the National Forests

logging, and California Senator S. I. Hayakawa was pushing legislation to curb more wilderness.

Shortly after the results of RARE II were announced, Senator Hatfield sponsored legislation to designate the 415,000 acres in Oregon that they recommended. The timber industry hoped that this legislation would include "release" language that would have removed the roadless tracts, which were not designated, from future forest planning processes and irrevocably dedicate them to logging. Conservationists fiercely fought this release language and were mindful that it could be embedded in override legislation.

While Congressman Weaver was a devoted environmentalist, he was also mindful that he represented a leading timber-producing district. Weaver was willing to champion a significant wilderness bill, but he was not willing to embrace a bill of the size sought by environmentalists. In 1982 and 1983, he put forth a bill for 1.1 million acres in Oregon, in contrast to the bill of 1.9 million acres proposed by environmentalists. And even that was less than the 2.9 million acres in the RARE II inventory. Weaver's bill (H.R. 1149) passed the House in March of 1983.

Other Oregon Democrats in the House helped Weaver and pushed specific units. Wyden pushed units around Mt. Hood: Salmon-Huckleberry and Badger Creek, as well as the North Fork of the John Day. And Les AuCoin provided additional leadership in the House. However, he had resisted having any units in his district; nonetheless, Sydney Herbert of the Lane County Audubon Society was instrumental in getting Drift Creek on the coast put in, with additional support from the Sierra Club.

Nothing moved in 1983 in the Senate. Hatfield wanted to control the situation, rather than splitting the difference with the House on Weaver's terms. However, ONRC and Audubon decided by December that they had to force action by filing their suit challenging Oregon's RARE II process. They also feared that the Reagan administration was about to begin to log the two million acres of roadless lands in Oregon they were not recommending.

Whether this suit forced Hatfield's hand is still debated. ONRC believes it finally forced him to bring forth a bill, while the Sierra Club believes that he had already indicated that his bill would come out in the spring. But in any event, in March of 1984, Hatfield got a slimmed down version of the House bill out of committee, and a few months later, it passed the Senate. It proposed designating 780,500 acres, and it included a version of "release" language that gave the timber industry far less than it wanted. It only barred reconsideration of the wilderness option during the current forest planning cycle. It could be reconsidered in the next one, and was called "soft release."

By embracing much of Weaver's bill, Hatfield avoided facing a difficult reconciliation process. Senate and House conferees soon reached agreement on a measure providing for designation of 861,500 acres as wilderness. It was called the Oregon Wilderness Act of 1984. It included Hatfield's "soft release" language.

Seventeen of the new units had significant amounts of old growth—most at lower elevations. Protection also came to vulnerable parts of the Cascade crest: with the Sky Lakes, Mt. Thielsen, and the Columbia units. Other existing wilderness areas along the crest were widened to include abutting forests: at Diamond Peak, the Three Sisters, Mt. Washington, and Mt. Jefferson. Additional areas, including Boulder Creek, some of the region known as the Middle Santiam, and Bull-of-the-Woods, now made it into the system.

Senator Hatfield felt exhausted by being forced to deal with two very controversial wilderness measures. Andy Kerr reports that Hatfield told him afterwards that this was his last wilderness bill. But circumstances did not turn out that way.

Opal Creek Wilderness—An Unusual Case

Controversy over protecting this unit—the Opal Creek unit north of Detroit Lake and east of Salem—was unusual in many ways. The controversy was disproportionate to its size; it was only 20,827 acres as finally set aside in 1996. It defied settlement for a long time—since the early 1970s. It had an atypical configuration; it ended up being a horseshoe-shaped wilderness around a core scenic area of 13,000 acres anchored by an old mining camp.

And its champion had a background more typical of a lumberman, whose first instinct was to try to get the Oregon legislature to save it as a state park (failed in 1989).

While the Forest Service owned most of the land there, some of it had been part of a mining operation run by the Shining Rock Mining Company, which sought silver, copper, lead, and zinc. It was started in the 1930s by James P. Hewitt and ceased operations in 1992. Hewitt's successors established ownership of ninety-five acres at Jawbone Flat in the midst of Opal Creek's old growth forests. His daughter had married Vic Atiyeh, Oregon's last Republican governor.

Vic's nephew George Atiyeh spent some of his formative years living there, which imbued him with a deep reverence for its "temples of nature." In his middle years, he went back to living there year round. While a lover of wilderness, he had a political life shaped by his family connections. He was a loyal Republican who had voted for Ronald Reagan, and he had even done some lobbying for timber companies.

Because he was attached to this beautiful remnant of old growth, he thought it ought to have a fate apart from the rest of Oregon's old growth. He had trouble accepting that local lumbermen just wanted to log his environs.

The Forest Service saw little of an exceptional nature in the Opal Creek drainage and simply wanted to remove most of its old growth. Even after its fate became enmeshed in contention, they planned to take most of it, with only a 500-yard scenic corridor to be left along the creek and the charming pool formed by the creek. They did agree to take it out at a slower rate. And Hatfield claims that he obtained a pledge from the Forest Service to delay logging while Congress had the matter under consideration.

The timber industry in that area was then reeling from depressed prices for timber and began to perceive this contest as a crucial test of strength. They feared that this was the opening wedge in a campaign to withdraw all old growth from cutting. They had no interest in finding a middle ground for compromise. Atiyeh tried to work out a compromise with an old colleague who was a local lumberman who had lived at Jawbone Flat. But the lumber-

Opal Creek

men would not budge. Hatfield said he saw this as "ground zero in the timber wars."

Twice Hatfield claimed he withdrew Opal Creek from omnibus bills for Oregon—in 1978 and 1984—at the request of Oregon's governor. One report has it that Governor Atiyeh changed his mind at the last minute, but that Hatfield resented his long resistance to wilderness designations elsewhere and was not about to reward him with this designation. But publicly Hatfield took the position that he would only move a wilderness proposal forward if there was a consensus to support it. One never emerged, including an effort at the end that he designed with a professional mediator. For six months, five environmentalists and five lumbermen negotiated in vain. Such a consensus protected a politician in a polarized situation, but it did not protect the environment.

CHAPTER 9—Breakthroughs on the National Forests

In 1994, the Democratic congressman from that district, Mike Kopetski, pushed a bill to protect Opal Creek through the House of Representatives. But Hatfield kept waiting for the consensus that would never come. ONRC mounted a vigorous campaign to budge him, with billboards in Portland rallying phone calls to his office. His lines were clogged for weeks.

George Atiyeh became embittered by Hatfield's obstinacy and felt betrayed. But he could not even get his uncle, who had been governor, to support him most of the time.

Finally, in 1996 as he prepared to wind up a long career in the Senate, Hatfield used his power and prestige to save Opal Creek. As chair of the Appropriations Committee, he was in a position in a conference on a defense appropriation bill to insist that the language he wanted to set aside Opal Creek as wilderness be included. It was a "going away present" of sorts for him. He claimed that he had always wanted to save it. No longer fearing a price he might have to pay at the polls for deserting the timber industry, he saved the wilderness around Opal Creek in the absence of a consensus.

2009 Wilderness Additions

The dynamics among Oregon's congressional delegation were different once Mark Hatfield left the Senate, particularly on wilderness designations. More inclusive processes were used to involve the interested public, and the members settled their differences through "give-and-take" negotiations. Partisan differences were also less evident. Also a weaker timber industry played a lesser role. All of these factors were especially evident in the package of wilderness additions enacted in 2009.

It had been twenty-five years since the last omnibus bill for Oregon wilderness was passed in 1984. Oregon Wild began this campaign as the new millennium was beginning, with additions to the various wilderness areas around Mt. Hood the center-piece of its campaign. But other areas were also involved and important to local citizens. The Oregon Natural Desert Association in Bend was focused on the case for the Oregon Badlands near there, as well as the Spring Basin unit—both of them BLM units. And anglers along the southern Oregon coast were intent on gaining wilderness protection for the Copper-Salmon unit in the Siskiyou National Forest. And other conservationists in the Ashland area wanted to complete the job in the Cascade-Siskiyou National Monument by setting aside the wilderness zone within it. Various bills were introduced on each by their congressmen: Peter DeFazio in the case of the Copper-Salmon area; Greg Walden for the two eastern Oregon units; and Earl Blumenauer for the Mt. Hood units. Senators Ron Wyden and Gordon Smith co-sponsored these bills in the senate. They were passed in 2006, 2007, and 2008, but failed to be reconciled because they faced a final obstacle in the Senate.

The obstacle in the Senate was the threat of a filibuster by Oklahoma Senator Tom Coburn, who was disgruntled over his inability to get the Senate to address his measure to drastically cut federal spending. But by 2009 the Democrats' numbers in the senate were sufficiently bolstered to enable them to finally muster the sixty votes needed to overcome the filibuster threat.

The Omnibus Public Land Management Act of 2009 brought two million acres into the national wilderness system in nine states. The leadership bundled up a whole suite of measures that had previously passed the House and Senate but were blocked by the filibuster threat. The wilderness title was only one of many in the measure. The omnibus bill at last passed the Senate by a margin of 77–21, and the House concurred by a vote of 285–140; in short order, the President signed the bill.

Senator Ron Wyden took the lead in pushing the Oregon bills, but sponsorship was soon bipartisan in nature. He negotiated readily with Senator Gordon Smith to reach agreement in light of public input at two well-attended public forums on the Mt. Hood units. Over 600 people participated in them, with over 100 community groups supporting the Mt. Hood package. The Mazamas were conspicuous in providing strong support. Thousands of constituents wrote supporting letters. The Oregon Natural Desert Association in Bend organized a coalition of over 200 groups to support the two units in the desert.

Five packages of wilderness bills had been worked out in Oregon and came to be embedded in the omnibus bill. They were for the Mt. Hood region, the Copper-Salmon area, the Soda Mountain Wilderness, and the two BLM areas east of Bend. Altogether, they added 213,000 acres of wilderness in Oregon.

Of this land, 140,300 acres were in national forests, and 72,200 were in BLM areas. Of the national forest additions, 127,600 acres were in the Mt. Hood National Forest, and 13,700 acres were in the Siskiyou National Forest. Of the BLM additions, 29,300 acres were in the Oregon Badlands, 20 miles southeast of Bend, and 6382 acres were in the Spring Basin unit overlooking the John Day River. Finally, 23,000 acres were in the Soda Mountain unit within the Cascade-Siskiyou National Monument of the BLM.

The Copper-Salmon Wilderness was particularly valued in protecting one of the most productive remaining habitats for salmon. It adjoined the existing Grassy Knob Wilderness on the east and lay along the headwaters of the Elk River and the Middle Fork of the Sixes River. Sixteen miles of old roads in it are being "put to bed" (i.e., removed and restored).

The Oregon Badlands wilderness encompasses a desert area of castle-like rock formations interspersed with juniper and serving as habitat for varied wildlife, including Yellow-Bellied Marmots, bobcats, pronghorn, and elk and deer. Some fifty miles of jeep trails among the volcanic pressure-ridges are now just walking trails.

The package for the Mt. Hood area established three new wilderness areas: the Lower White River Wilderness on the southeast flank of Mt. Hood, the Roaring River Wilderness on the south side of the Salmon-Huckleberry Wilderness, and the Clackamas Wilderness. It made additions to already established units: the Salmon-Huckleberry, Badger Creek, Mark O. Hatfield, Bull-of-the-Woods, and Mt. Hood.

The legislation also added rivers in a number of places to the national system of Scenic and Wild Rivers: eighty miles on nine rivers in the Mt. Hood National Forest and 9.3 miles on the North Fork of the Elk River in the Siskiyou National Forest.

In addition, the legislation set up a new 35,000-acre National Recreation Area (NRA) on the south side of Mt. Hood around Government Camp, as well as establishing various areas to be given special management. And it consummated the exchange of private lands around Cooper Spur, which had been the site of a controversial proposal for expanding a ski area.

This time the obstacles confronting wilderness promoters were not in Oregon but elsewhere. And the new areas continued the trend of making additions that favored habitat, adding fragments of wild land, and new types of areas. The old types of areas at high elevations have been largely taken care of. And a less contentious, more inclusive and responsive process, has been embraced. Let us hope that it heralds a new day.

At the time this is written, there were forty-seven units in the National Wilderness Preservation System in Oregon, totaling 2.4 million acres.

References

Alan Tautges, "The Oregon Omnibus Wilderness Act of 1978, Public Law 95-237," *Environmental Review* (Spring 1989).

Dennis R. Roth, *The Wilderness Movement and the National Forests* (Intaglio Press, College Station, Texas, 1988).

Andy Kerr, *Oregon Wild: Endangered Forest Wilderness* (Oregon Natural Resources Council, Portland, Oregon, 2004).

Doug Scott, *The Enduring Wilderness* (Fulcrum Publishing, Golden, Colorado, 2004).

Gerald W. Williams, *The U.S. Forest Service in the Pacific Northwest* (Oregon State University Press, Corvallis, 2009).

CHAPTER 10
IMPORTANT FEDERAL INITIATIVES AFFECTING OREGON

Northwest Power Act

A clutch of intersecting concerns in the late 1970s led to the enactment in 1980 of the Northwest Power Act. It has set the context ever since for the interplay of issues governing the production of hydro-power, salmon survival, and meeting demand through improving efficiency.

By law it established new goals for the region that require balance in planning so that equal attention is given to the needs of fish and power supply. It also requires that appropriate efforts be made to restore salmon numbers and mitigate losses and sets up a new planning council for the Northwest. It requires them to use the best available science.

Among its purposes, the law directs the council to assure that the future power supply is both efficiently derived as well as economically produced; it also needs to be adequate and reliable. And it put the Bonneville Power Administration (BPA) in a pivotal position in managing this process.

As background, the main players had been locked in increasing conflict throughout the 1970s over these issues. On one hand, the power forces had been warning of likely shortages in our future and wanted new projects to be brought on line, particularly nuclear plants. In light of the anticipated shortages, BPA had been warning public utilities that it could no longer even meet their demands in fulfillment of their preferential rights. And public power utilities were facing a program in Oregon to create a statewide public utility to get BPA to distribute power to rate payers served by private utilities.

On the other hand, environmentalists had been arguing more and more skillfully that improvements in efficiency could obviate the need for new power plants. Plans for a scheme of new nuclear plants in Washington state (WPPSS) had just collapsed. And the tribes had just won an important lawsuit challenging the impact of federal Columbia River dams on tribal treaty fishing rights.

The Congress acted when the conflict became too acute, with Senator Mark Hatfield guiding the way. Some of the key language was drafted by Larry Williams when he was at the Council on Environmental Quality under President Carter. It broadened BPA's mandate to embrace conservation and efficiency, not just maximizing power production. Similar language had earlier been drafted by Jim Blomquist of the Sierra Club's northwest office in the mid-1970s when the Club had tried to reform BPA's basic law of 1937.

The planning council is composed of two representatives from each state in the Northwest: Washington, Oregon, Idaho and Montana. The governors appoint these members. Each state legislature had to approve setting up this council on this basis. It is argued that this arrangement gives the people in these states a stronger voice in setting the terms under which the states' common resources will be managed.

The council approves plans for twenty-year periods, which it must bring up to date every five years. For instance, in 2010, they proposed meeting future needs largely through more efficient energy use. The council also must approve programs to restore and protect fish and wildlife adversely affected by the dams on the Northwest's rivers. All of the federal agencies involved in building and operating dams in the region are also involved in implementing these plans.

In addition, by virtue of the key lawsuit they won, the tribes have been accepted as a serious player. BPA has been required to fund a new inter-tribal entity, the Columbia River Inter-Tribal Fish Commission, which helps to restore salmon. BPA also must take steps to restore salmon runs by releasing more water from dams and investing money in restoring habitat. While it is doing these things, controversy continues over whether they are making much of a difference.

BPA has the unenviable task of trying to balance and reconcile goals in its guidelines that sound almost all of themes of the region's competing interests. It remains to be seen how well they play the role of "philosopher kings" who can divine how this may be done. Some think they have not been able to fulfill their promise.

Spotted Owl Reserves

In due course, the timber wars in the West came to focus on old growth and their use by the Northern Spotted Owl. But the wars had raged across western forest regions through the 1970s and 1980s. In the Northwest, the permissible cutting levels in federal forests in this time had more than doubled. Some thought that this was merely conflict between a long-dominant industry and a new challenger: the environmental movement.

Others in this movement saw it as a changing of the guard in that movement between the old recreational conservationists and the "naturalist" contingent more interested in the welfare of wildlife, especially the less charismatic species. And this was a time when changes in the laws brought many more biologists into the workforce of the land management agencies. Information was moving from these biologists into the hands of citizen activists.

While groups such as the Sierra Club had long been locked in conflicts with the timber industry, they had never dared to try to shut it down. The bruising conflicts that they had experienced taught them to believe that this was "political suicide"—i.e., such attempts would put an end to their ability to influence political events in the Northwest. The newer contingent had not been through these experiences. Some were groups such as the Oregon Natural Resources Council, and others were groups such as Audubon Societies that rarely had been on the barricades. Now they saw the need to close large areas of old growth to logging.

And this was at a time that the results of over-cutting were beginning to catch up with the timber industry in the northwest. Most private stands had been liquidated by then, with as many as 200 mills closing. Public timber provided the remaining supply, mainly in the national forests. As Oregon's leading congressmen pushed the Forest Service to sell even more on these forests, the logging was yielding less as the prime stands were being exhausted (i.e., having fewer board-feet per acre).

But many argued that closing these stands to sale would stagger this industry. Already greater mechanization in the 1980s had led to major reductions in the jobs in the industry: 25,000 jobs had been lost in this period. Moreover, the decline of jobs in the timber industry had been continuing for decades.

Furthermore, in the 1970s a new guiding law for the national forests had been enacted: the National Forest Management Act of 1976. Its provisions required the Forest Service to maintain the health of forest ecosystems and to maintain populations of vertebrates in them so that they did not fall below minimum viable levels. More specifically, it charged them with identifying what biologists called "indicator" species and with protecting them.

The Endangered Species Act that had also been enacted a few years before that (1973) required the Fish and Wildlife Service to list species that are endangered or threatened. It also prohibited "takings" of such species and required that agency to prepare plans for their recovery. Habitat destruction was regarded as a kind of "taking." And, of course, the National Environmental Policy Act (NEPA, 1970) required that federal agencies prepare environmental impact statements whenever their actions might cause significant adverse impacts to the environment.

One of the Forest Service's new biologists, Eric Forsman, in the early 1970s found that a species in the Northwest was the victim of shrinking areas of old growth. The numbers of Northern Spotted Owls, which lived in these forests, were declining rapidly as the old growth acreage shrank. He found that each pair of these owls needed between 1200 and 3000 acres of old growth as its habitat. Because its welfare depended on the existence of old growth, its shrinkage indicated what was likely to be happening to other species that lived there. This status made it an "indicator" species. It was a vertebrate whose numbers were in danger of falling below minimum viable numbers. A task force also concluded that a minimum of 400 pairs were needed for the viability of the species.

At this point, environmentalists unleashed a barrage of legal attacks. They began with a series of administrative appeals in Eugene over BLM's old growth logging plans. They were triggered by the Lane County Audubon Society's Sydney Herbert, who had just put on a pivotal conference with the involved agencies (see page 183). She enlisted the aid of the Law School's John Bonine and students in its law clinic. They filed a series of administrative appeals of the agency's sale plans, which they characterized as "owl reduction" schemes. A similar appeal was launched of Forest Service timber sale plans. Filed in 1980, they were the first official administrative appeals filed to protect spotted owls. While they were rejected, more were filed in 1983, this time after the BLM had executed a working agreement on their logging with the Oregon Department of Fish and Wildlife. The clinic built its appeal around the absence of an EIS on these agreements. This appeal was on behalf of various local environmental groups: the Lane County Audubon Society, ONRC, and the Oregon chapter of the Sierra Club.

While this was also rejected, the BLM then did an assessment to determine whether there was any new information. Concluding there was not, they rejected this appeal too. Thereupon, the clinic filed a lawsuit in December of 1987 against the Interior Department on the EIS issue. It was the first suit to be brought to defend old growth and the spotted owl.

These legal efforts on behalf of these owls and forests outraged the state's timber interests. They, in turn, unleashed attacks upon the law school's legal clinic, feeling it should not accept clients that attacked the state's leading industry. Because Sydney Herbert had encountered difficulties in finding representation, Bonine felt that he should teach his students to even accept clients with unpopular causes. He and fellow professor Michael Axline did see the case through the process. In light of the backlash, the university president had to appoint an independent committee to review their actions, which stood by them. However, at an earlier time (1982), the industry did succeed in forcing out a lawyer attached to the clinic who worked for the National Wildlife Federation.

Other environmentalists in the Northwest faced the fact that the Fish and Wildlife Service was unwilling to list the Northern Spotted Owl as threatened—a designation applied to species that, in the foreseeable future, were likely to become endangered throughout all or part of their range. In January of 1987, thirty environmental groups had petitioned to list it as endangered. Notwithstanding all of the new scientific findings, the Fish and Wildlife Service refused to budge.

For this reason, the Seattle Audubon Society and twenty-one other like-minded groups brought suit in federal court in the spring of 1988 to challenge this decision under the terms of the Endangered Species Act (ESA).[33] The suit was brought by the Seattle office of the Sierra Club Legal Defense Fund, which by then was independent of the Sierra Club. They charged that this refusal was arbitrary and capricious and did not comport with the best available science.

The judge in the case (Thomas S. Zilly) quickly directed the agency to reconsider its position, finding that expert opinion was contrary to the position it was taking.

In the spring of 1989, the Fish and Wildlife Service (now with new political oversight) changed its position to state that it was now considering the species to be threatened. In response, the Forest Service began to take action to substantially reduce its timber sales in Northwest forests.

Shortly after the suit and two others had been filed, the Forest Service set up an Inter-Agency Scientific Committee to prepare a strategy to safeguard the owl. It was chaired by Jack Ward Thomas, a Forest Service career biologist. It soon issued a report recommending that portions of the old growth

---

[33] In addition to many of the plaintiffs from the national scene and from Washington state, they included eleven from Oregon, among them ONRC and many local Audubon Societies.

be set aside across the Northwest to provide Habitat Conservation Areas for the owl. These would reduce the risk of the species becoming extinct. Just as this report was released, the Fish and Wildlife Service issued its official finding that the species was classified as threatened.

Then in May of 1991, in response to another suit by Seattle Audubon, Judge William Dwyer in Seattle issued an injunction holding up all timber sales affecting national forest owl habitat in the Northwest until the Forest Service complied with the provisions of the National Forest Management Act that required it to maintain conditions needed by indicator species, such as the Northern Spotted Owl. Dwyer issued a judgment condemning the "deliberate and systematic refusal of the Forest Service" to comply with laws protecting wildlife. He made it clear that he was not talking about the agency's scientists but was referring to higher authorities. In context, he was talking about the political appointees in the executive branch. In upholding his decision, the Court of Appeals made it clear that the Forest Service was required to both comply with the diversity requirements of NFMA as well as protective requirements of the ESA.

At the behest of the House of Representatives, the agencies managing federal lands convened a team of four scientists that mapped out the public land areas in western Oregon and Washington used by species dependent on old growth. Various options were laid out for rescuing these populations of dependent species. When their report was delivered to Congress that summer, the Congress was too polarized to act.

It was only when Bill Clinton was elected President the next fall that the Executive Branch took decisive action to address these problems. By now there were over a dozen lawsuits[34] and three court injunctions involving old growth logging, virtually closing down sales of timber in old growth. In April of 1993, he convened his "Forest Summit" in Portland at which experts laid out their findings and the options, with invited stakeholders responding. Observing were five cabinet secretaries, as well as Clinton and Vice President Gore.

---

34 Some of them were spearheaded by the Portland Audubon Society.

By that summer, Clinton announced he had chosen Option 9. Clinton said that this option would comply with the law to protect impacted wildlife, end courtroom gridlock, and provide an assured supply of logs to rural logging towns.

Clinton's plan was called the Northwest Forest Plan and was applied to 25 million acres in federal forests in the West Coast states[35] Of this, thirty percent were old growth forests that had been open to regular logging, which were called "late successional reserves," comprising 7.4 million acres—of which 5.6 million acres were in Oregon. But these are interspersed with young growth forests that had been clear-cut; less than half of this acreage displayed classic old growth characteristics. According to definitions in use then, old growth forests have a high incidence of large trees (six trees per acre of at least 200 years of age under an earlier definition), an average age now in excess of 180 years, with layers of tree tops and canopies with a high degree of closure, as well as snags, down logs, and rotting wood. They are the last stage in forest succession here, but with considerable variation in characteristics.

Almost half of the plan forests were in areas that had already been reserved either administratively or by Congress, or were in strips called "riparian reserves." Some twenty percent were in areas reserved for careful timbering ("matrix" lands) or put in experimental "adaptive management areas." In 1994, Judge Dwyer accepted it as meeting the terms of the National Forest Management Act.

Option 9 put most of the remaining blocks of old growth timber in late successional reserves that would be off-limits to logging that changed its character or structure. It also applied to BLM's holdings on the O & C lands. Seventy percent of the old growth areas were put in these new reserves; the remaining thirty percent remained open to regular sales. Over time, an

---

35 In official jargon, these lands were allocated as follows: 30% were put in late successional reserves, 16% were put in matrix areas open to restricted logging, 11% were put in riparian reserves, 6% were put in adaptive management areas, 1% were in managed late successional reserves, 6% fell in areas already administratively withdrawn, and 30% in areas already withdrawn by Congress.

Chapter 10—Important Federal Initiatives Affecting Oregon

effort would also be made to restore old growth characteristics of the interspersed areas that had been logged.

As a consequence, the volume of timber sales in northwest public forests dropped to one-fourth of its peak levels in the 1980s.[36] Only about one percent of the older forests still open to logging were touched in the decade following Clinton's order—about 12,000 acres in Oregon. While about 17,000 jobs were lost in the woods products industry, this number was only about one-third of that forecast by the industry.

In 2006, the Bush Administration tried to take O & C lands out from underneath their status as late successional reserves and to triple the volume of timber sales under what was known as the WOPR (Western Oregon Planning Revision) program. Environmentalists protested, with the Obama administration reversing this decision. But then the timber industry protested and sued. Then in turn the environmentalists went to court (2011) and obtained a ruling in 2012 that the protective standards of the Northwest Forest Plan were still in force on these lands, vacating WOPR.

Congress has never explicitly blessed these late successional reserves. But the Fish and Wildlife Service has been gradually expanding the amount of acreage it has designated in the Northwest as critical habitat for the spotted owl, adding four million more acres in Oregon in 2012. But at least in the national forests, these reserves have survived for nearly twenty years. The longer they survive through changing administrations with contrasting views, the more likely it is that they will become permanent. Probably the peril of the O & C forest lands has been a trial balloon.

Their chief peril now, however, is that the viability of the Northern Spotted Owl is threatened by an owl invading from the east—the Barred Owl. It tends to displace Northern Spotted Owls. The Fish and Wildlife Service is now proposing that their numbers be reduced by shooting them—releasing more space for spotted owls.

---

36 Between 1989 and 2000, the amount of timber removed from federal lands in Oregon dropped by ninety-six percent as a result of various factors: the NW Forest Plan, over-cutting, and more lawsuits (source: OR Forest Resources Institute). In 2011, only a half-billion board feet were cut.

We are also learning that there are other varied species that depend on old growth habitat (about a thousand of them), such as the Red Tree Vole and the Marbled Murrelet. The Portland Audubon Society has taken the lead in having a large area in the Coast Range identified as an Important Bird Area for Marbled Murrelet habitat. The needs of these and other species may help justify retaining these late successional reserves. The courts have held that the viability requirement under NFMA extends to the entire biological community and is not limited to the indicator species. Also the importance of these other species has grown as the "Survey and Manage" requirement has been reinstated by the courts. It requires that a careful look be taken for diverse species on a site before it is logged and their viability be protected, if found. And new research by OSU forest scientists suggests the value these forests have in sequestering carbon. Time will tell whether the late successional reserves will survive or succumb.

But already it would appear that the long-term orientation of the agencies managing federal forest lands is changing away from commercial timber production toward preserving biodiversity on their holdings as a result of the Northwest Forest Plan, as well as other factors. This is an historic achievement. Many national forests that once had been big timber producers now are merely thinning dense stands of young growth and removing surplus logging roads. They also are focusing more on selling special forest products, such as mushrooms and products sought by the florist trade. Researchers Jerry Franklin and Chris Maser have played notable roles in moving federal forestry agencies toward eco-system management.

It is amazing too that an industry that once was as powerful as the timber industry has faded so quickly. Perhaps its frantic resistance at Opal Creek portended its change of fortunes.

[See the list of relevant lawsuits that triggered the Northwest Forest Plan, or arose out of the matter, following the list of references at the end of this chapter.]

Chapter 10—Important Federal Initiatives Affecting Oregon

Courts Manage Salmon

Salmon have been an object of controversy for generations. As they began their decline in 1880, everyone has put forth arguments for remedies. They have now have been locked in litigation for decades.

But notwithstanding the arguments over who gets to take them and how much and over all the uses that affect their habitat, they remain in deep trouble. Half of the wild populations in the Columbia basin are now extinct. Only five percent of the original wild salmon remain. Only 400,000 wild ones still spawn out of the estimated 16 million that once did.

The National Oceanic and Atmospheric Administration (NOAA) now regards thirteen runs of northwest salmon as endangered. Throughout the 1990s, it listed runs as endangered, beginning in 1991 and again in 1998–1999.

The proliferation of dams on the Columbia and its tributaries has limited these runs more than any other factor, but other factors have also contributed. Habitat has been degraded along tributaries, streams, and wetlands. They have been too heavily fished, despite regulations limiting take. Predators are inhibiting recovery of a species subjected to too many stresses.

When the dams were being built, fish raised in hatcheries were viewed as an alternate way to produce salmon in abundant numbers. And 2.2 million hatchery fish have returned in a recent year (2009). But recent science shows us that hatchery fish lack the robustness of native stocks and through interbreeding weaken the wild stocks over time (they spread hatchery diseases and are less prolific reproducers). They are now seen as part of the problem, although hatchery admirers point out that wild fish in streams without hatcheries are doing as poorly as those in streams with hatcheries.

But the key question is whether the stocks can recover just by releasing more water over the dams and restoring habitat. Most Northwest politicians want to believe that the dams and salmon can co-exist in the Columbia basin. They have been pressing NOAA and BPA to take this position. But the scientists in the American Fisheries Society believe that some of the barriers produced by these dams need to be removed or breached. They believe this is particularly true of the four dams on the lower Snake River. Biologists in Oregon's Department of Fish and Wildlife concur.

Fourteen federal dams on the Columbia system have done most of the damage: six on the main stem, four on the Snake, and four more on tributaries in Idaho and Montana. And three dams block upstream fish passage entirely: Grand Coulee, Chief Joseph, and Idaho Power's projects on the Snake in Hells Canyon.

All told, 147 stream miles have been turned into reservoirs that subject downstream migrants to a procession of perils. Reservoirs at times have insufficient oxygen, and these migrating stocks must navigate slack waters over a longer time, during which they are exposed to predators such as pikeminnows. When smolts (juveniles ready to migrate) are drawn into turbines at the dams, they can be killed (by pressure changes or blade strike) or become disoriented on coming through, becoming ready prey for predators. Or they can be swept into churning water with dangerous levels of nitrogen (causing a sensation similar to "the bends"). To avoid these perils, sometimes they are barged or trucked around the dams. When they are not, efforts are made to screen the smolts out of the turbines and guide them into fishways to bypass the turbines and turbulent water.

NOAA has been trying to get its biological opinions approved since 1992, but its opinions have repeatedly been challenged in court by anglers and conservationists, the state of Oregon, and at least one tribe. Six opinions have been rejected by judges. Under the Endangered Species Act, the managing agencies embody their technical findings in their recovery plans and recommendations in something called a biological opinion.

From 2001 to 2011, Judge James Redden of the federal district court in Portland was in charge of adjudicating the matter. In 2001, the National Wildlife Federation challenged a ten-year recovery plan proposed by NOAA. In 2004, he found it deficient in failing to take enough steps to assure recovery of the listed salmon species. He found further revisions deficient again in 2005 and 2008 and ordered more water spilled from the dams to induce salmon to migrate upstream across these barriers. He also directed them to assess the option of breaching the Snake River dams.

CHAPTER 10—Important Federal Initiatives Affecting Oregon

At the time that this is being written, NOAA is under the oversight of the Obama administration and is being directed by Dr. Jane Lubchenco (2009–2013), a marine biologist from Oregon State University. Conservationists fear that she has fallen afoul of the political constraints of being in a high appointed position and is not willing to face up to dealing with the stark realities. They question whether she is speaking entirely from the vantage point of scientific expertise.

Her defenders point to the record: in the last thirty years BPA has spent $3 billion on restoration, hatcheries, and research. Five thousand acres of denuded bank-sides have been restored, as well as 1000 acres of estuary flood plain. Much more water is being spilled over the dams at critical times. In some cases, such as in the Umatilla, more water is being left in the river for fish because Congress funded a scheme in 1992 to provide an alternate source of irrigation water by pumping water from the Columbia down a canal to replace the river water they used to take.

In the last fifteen years, the Corps of Engineers has spent $1.3 billion on facilities on its dams to improve passage of fish and to provide for their protection. Devices called sluices and slides are being installed, which supposedly will enable ninety-five percent of juvenile fish to successfully pass downstream through dams. Five million fry are now being released from hatcheries.

Runs of salmon above Bonneville Dam have markedly improved since 2000, subject to annual fluctuations. Runs of spring sockeye and fall chinook on the Snake have also improved. While admitting that improved conditions in the ocean have played a part, they point to the actions they have taken (i.e., habitat restoration and greater releases of water) as producing results.

But the critics are not convinced. For half of the endangered populations, productivity is too low for the fish to recover. The agencies have fallen behind the targets they gave the judge: on the upper Columbia and on the Salmon River, they are achieving only twenty percent of the needed levels of recovery; on the lower Columbia, they are reaching only twenty-five percent of their targeted amount of recovery; on the main stem above Bonneville, recovery levels are at only forty percent of the targeted levels. On tributaries, recovery is lagging in half the cases.

In light of these lagging results, the critics believe that promises of salmon recovery under current plans are entirely too speculative.

On the other hand, the agencies feel that they will soon (by 2013) be able to get their recovery work on target—in light of the contracts for work that they have signed.

In his third opinion, in August of 2011, Judge Redden found that NOAA's new biological opinion was specific enough through 2013, but not thereafter. He ordered them to produce a more specific plan by the end of 2013. He further required that more water be released at key times. And he also called for more information on breaching the dams on the lower Snake River.

Judge Redden said his court would continue to retain jurisdiction over this matter, but ultimately the wild fish will indicate who is right. Now at age 83, he has retired and turned this matter over to his successor. Few judges have had such a consequential matter in their hands for so long, having had a profound impact.

He has given the fish a better chance of making it back and has fostered a deep-seated respect for the courts. He has shown a tenacity that is quite unusual. He has earned our respect and gratitude.

References

William G. Robbins, *Landscapes of Conflict* (University of Washington Press, Seattle, 2004).

Gerald W. Williams, *The U.S. Forest Service in the Pacific Northwest* (Oregon State University Press, Corvallis, 2009).

Scott Learn [series on salmon], *The Oregonian* (May 8–9, 2011)

Also online sources.

Re the Spotted Owl litigation, these are the leading cases:

Northern Spotted Owl v. Hodel, 716 F. Supp. 479 (W.D. Wash. 1988) and *Northern Spotted Owl v. Lujan,* 758 F. Supp. 621 (W.D. Wash. 1991).

Seattle Audubon Society v. Robertson (1990) and *Robertson v. Seattle Audubon Society* (1990), 503 U.S. 429.

Seattle Audubon Society v. Evans, 771 F. Supp. 1081 (W.D. Wash. 1991); affirmed on appeal in *Seattle Audubon Society v. Evans,* 952 F. 2d 297 (9th Cir. 1991).

Seattle Audubon Society v. Moseley, 798 F. Supp. 1473 (W.D. Wash. 1992).

Seattle Audubon Society v. Lyons, 871 F. Supp. 1291 (W.D. Wash. 1994); affirmed on appeal in *Seattle Audubon Society v. Lyons,* 80 F. Supp. 1401 (9th Cir. 1996).

Portland Audubon Society v. Lujan, 884 F. 2d 1233 (9th Cir. 1989), and *Portland Audubon Society v. Lujan,* 109 S, Ct. 1470.

CHAPTER 11
CONCLUSIONS

Persisting Progress

Conservation has become a habit in Oregon. Breakthroughs have been made since 1886, and they keep happening in every decade. They are not just the product of the politics of a given time, or a swing in the political pendulum. Certainly, conservation seems to prosper under progressive auspices, but those promoting it just seem to find a way to keep moving relentlessly forward.

This record is set forth graphically in Appendix A—labeled "Timeline of Conservation Accomplishments in Oregon." During more conservative times, progress is made through administrative actions or via initiative measures. Yes, the pace slows in the 1920s, but it picks up soon afterwards. And the first anti-pollution effort, which can be viewed as an environmental measure, as well as a conservation one, commences in the 1930s. And this effort to clean up pollution in our rivers is sustained thereafter.

Other types of citizen action for conservation reappear right after the end of World War II. While public unrest grows in the 1960s, it begins in the mid-1930s and continues in the mid-1940s. And the scope of environmental activism intensifies by the mid-1960s, at the same time as in other West Coast states, which think of their expanded consciousness as uniquely early (i.e., before Earth Day in 1970).

Someone, or some entity, always seems to step forward to lead when the times call for it. Sometime a leader emerges from the shadows; sometimes an ordinary citizen becomes energized over something that is just not right; at times a quiescent organization comes alive; on occasion a charismatic politician rallies the public; and there are cases when a bureaucrat suddenly becomes an enlightened leader. Once engaged, they just do not give up. They find ways to arouse or rally the public.

A Place of Importance

Oregon has stood out as a pivotal place for the resolution of key issues. It has been ground zero in shaping how laws and policies would be worked out for the national forests. It was the center of efforts to combat fraudulent public land claims.

And Oregonians have learned how to work within the political system to achieve their ends, as activists have persuaded presidents to withdraw lands from settlement for Crater Lake National Park and have gotten federal wildlife refuges set aside. They have learned how to mobilize public opinion through changing ages.

Innovation

This historical record not only indicates steady progress, it also reveals impressive innovation. Oregon has not only been a wellspring of innovation in broader public policy, it has also been one in the field of conservation and the environment. From the outset, citizens in Oregon have asserted their interest in seeing national forests serve broad public interests. They have been among the earliest in pressing to have waterfowl protected in refuges and in combating wanton slaughter. It was a pioneer in shaping laws to guarantee the public access to beaches. At an early date, they tried to find ways to protect forest scenery along highways.

They were among the first to build a scenic highway, as at Columbia Gorge, that sought to reconcile the tension between construction and the environment. And Timberline Lodge was not only notable for the skillful way it is blended in its mountain setting, it was the first park-like lodge that was owned by the public rather than a concessionaire. Oregonians also embraced the goal of restoration at an early time when they set up Portland's Forest Park, the country's largest city park that aspired to be wild as well.

In modern times, Oregonians have innovated even more. In the field of conservation, Oregon broke new ground with the formula developed for the Columbia Gorge Scenic Area. It was unique in a number of respects: in its size, in the way that it combined development and preservation, and in its emphasis on planning and regulation. The Cascade-Siskiyou National Monument forged new ground in being based just on preserving biological diversity and not on other values. Eugene's wetlands forged new ground in a number of respects: restoration, as well as biological diversity, and a consensus process.

In the legislation to protect Steens Mountain, the formula is unusual, but the most innovative feature is the Redband Trout Reserve, which is the first use of this formula to protect an endemic fish. The approach taken in the legislation establishing the Newberry National Volcanic Monument is innovative in designating a Special Management Area, which can be administratively added to the monument later if no development (i.e., drilling) takes place there.

Inasmuch as Oregon is the central focus of the Northwest Forest Plan (though other states were involved), it is worth mentioning that this innovation embodies a move toward establishing a system of reserves that focus on the structure and function of the forest rather than just preservation. And it has played a major role in re-orienting the Forest Service away from timbering and toward ecological management.

Even wilderness expansions in Oregon have moved in new, innovative directions. Now they try to protect habitat for wildlife and fish, as well as fragments of wild land (rather than just cohesive blocks in the high country). And in the way that Oregon has led in placing less threatened rivers and streams in the federal system of Wild and Scenic Rivers, it has innovated in finding another way to preserve a lot of acreage, in this case along stream banks.

Oregon's Department of Fish and Wildlife has shown leadership in opposing the NOAA's tepid recovery plans for salmon and letting its biologists collaborate with other fishery scientists in the call for the removal of lower Snake River dams to restore salmon runs. And Portland has led the way in fostering a program for its significant trees that actually protects them.

In the environmental arena, Oregon has innovated with its bottle bill, with its statewide program for land-use planning standards and regulation, and with its early experiment in establishing a greenway. In later years, it has been among the leaders in turning away from problematic sources of electric power—nuclear and coal, and establishing mandates for a growing component of renewable power.

Certainly, other states, such as California, have been important innovators too, but in relation to the size of its population, Oregon stands out. On a per capita basis of innovation, Oregon is a leader.

On the same basis, Oregon is notable for the variety of its formulas to protect land and the environment. With regard to reserves, it has national forests, national grasslands, national parks, state parks, national recreation areas, natural areas,[37] marine reserves, private reserves, areas of critical environmental concern, outstanding natural areas, a cooperative [conservation] management area, national scenic areas, scenic and research areas, wilderness, wild rivers, trout reserves, estuarine reserves, and national monuments to protect: scenery, caves, bio-diversity, paleontology, and volcanic phenomena. And now it has reserves in the making with its late successional, old growth reserves.

In the environmental arena, it tries to curb pollution, protect groundwater, move toward safe power, plan land use, reduce litter, take back discarded electronics, and control surface mining.

It is almost alone in the varied ways it has sought to protect the environment and foster conservation. Many have involved the state's portion

---
37 In Oregon, some call them Natural Heritage Conservation Areas

of broader federal programs, but others have involved programs under the auspices of the state government, and local government.

Political Parties and Conservation

In thinking about how we accomplished all this, one might think it would be easy to characterize the respective roles of the two main political parties. At the time this is written, it seems to be straightforward: most Democrats support environmental protection, while most Republicans do not. But that has hardly been the record over the time reviewed in this book. At times, Republican presidents, such as Theodore Roosevelt, have been zealous champions of conservation, while President Wilson had a lackluster record. While President Franklin Roosevelt did many things for the environment, he also began the dam-building era that destroyed the salmon runs. Even Teddy Roosevelt promoted irrigation projects that drained the wetlands of many wildlife refuges.

In Oregon itself, John B. Waldo, William G. Steel, and Ben Olcott were Republicans—albeit progressive ones. Os West was a progressive Democrat who kept the beaches open for the public and fought land fraud. But Walter Pierce and Charles Martin were Democratic governors of a different stripe who showed little interest in conservation. While Republican Tom McCall was a charismatic proponent of environmentalism, low-key Bob Straub, a Democratic governor who followed him, had his own ways of stirring up public interest in environmental ideals. Both had good environmental records, with McCall credited with getting a hundred environmental measures through. But neither of them was willing to do much to antagonize the timber industry. Neither of them had a perfect environmental record.

While Richard Neuberger showed signs that he might become an environmental champion, he died (in 1960) before that era dawned. During his time, he tried to be on every side of the salmon issue, sometimes opposing habitat-killing dams and sometimes promoting them.

Chapter 11—Conclusions

Wayne Morse is not easy to characterize since over time he belonged to all major parties, but he is best known as a strong liberal. Yet he formed his outlook before environmentalism had much of a following. He helped get funding for most of the federal dams built in Oregon that ruined salmon runs and assiduously sought federal funding for timber access roads in the national forests that facilitated wiping out most of the old growth. During Mark Hatfield's long career in the Senate, he trod the same path, just as they did when they were both Republicans. Hatfield was a centrist who delivered on the things he thought that voters wanted.

Today Democrat Ron Wyden sometimes champions wilderness, but at other times wants to revive the timber industry. Jeff Merkley has a record that is promising—being the only senator with a perfect voting score (according to the League of Conservation Voters)—but has yet to chalk up notable accomplishments. In recent memory, Democratic congressman Jim Weaver was the most committed environmental champion, but there were limits to what he would do. Peter DeFazio has a solid environmental record, but there were limits to what he would do. Les AuCoin pursued a cautious, middling course, often pushing bills promoting the timber industry just as had Hatfield and Morse.

Democratic governor John Kitzhaber deserves to be noted for the way that, in his first term, he courageously vetoed the efforts by Republican legislators to roll back the rigor of the state's land use laws. Ted Kulongoski stands out because he championed a suite of energy measures emphasizing renewables and efficiency, and protection of the Metolius, and ushered in a program where state parks were created again (though most emphasized recreation facilities). Barbara Roberts has also taken good stands, but has not stood out in our history. Environmentalists remember Neil Goldschmidt more for his years as Portland's mayor than his years as governor, though he initiated an intriguing benchmarking process.

While Republicans tend to want to protect business from the heavy hand of environmental regulation, their distaste today for government causes them to do little to foster government programs. On the other hand, while Democrats usually favor stronger regulation, they also often want to foster government programs that often destroy the environment (though less so in recent years). Perhaps this tendency characterized moderate progressives such as Hatfield as well. And when the Oregon legislature adopted its mandate for renewables, Republicans were among those who supported it.

Perhaps it is confusing to equate progressive politics with environmentalism. While some think progressive accomplishments in Oregon crested in 1978, environmental accomplishments have marched on since that time. They do not seem to have marched to the same drummer.

The future may chart a clearer path for the role of the two parties in treating the environment, but the path in the past is anything but clear.

Failures

While the record is rich with accomplishments, it is also replete with failures. Oregon is hardly an ecological utopia.

It has lost ninety percent of its old growth, and ninety-five percent of its wild salmon. Nineteen species have gone extinct in Oregon, and 200 are quite vulnerable. It allowed over four million acres of its state forest lands to be pirated from its hands, as well as too many of its tidelands. Nearly seventy percent of its coastal tidelands were then lost to diking and draining. Overall, thirty-eight percent of its original wetlands were lost, including nearly half of those in the Willamette Valley.

The canyons of the middle Deschutes and Crooked River were lost to PGE's Pelton and Round Butte Dams, along with the salmon habitat there; Sam Boardman had called their confluence "the Grand Canyon of Oregon." He lamented the coming incursion into Cove Palisades State Park. Oregon politicians and government did put up some resistance in the mid-fifties, but it was not enough. No effort was made in Oregon to save Celilo Falls, which was flooded out in 1957 by The Dalles Dam (though technically, the dam is on the Washington side of the river). It had been one of the nation's most productive Indian fishing sites.

Chapter 11—Conclusions

Curry County Coastline

It has failed to rally behind efforts to create national park units along the coast in Curry County (as the National Park Service wanted in the 1930s),[38] or in the central Oregon Cascades, or around Mt. Hood or at Hells Canyon as some conservationists have wished.

The jury is still out with regard to establishing reserves in other areas awaiting recognition for their special values: the awesome Owyhee Canyons in southeastern Oregon (which have two million acres of wilderness and are nearly 1400 feet deep) and the treasure-house of biodiversity in the Siskiyou region of southwestern Oregon, which the World Wildlife Fund has called "globally outstanding." The future will determine whether the people of Oregon recognize what they have to offer.

And we have to recognize that non-commodity values have received little emphasis in the management of state lands, such as the Tillamook and Clatsop State Forests. These forests harbor over seventy species of wildlife

---

38 In 1936 the National Park Service proposed creating a 30,000-acre park on the southern Oregon coast between the Pistol River and Brookings; Senator McNary introduced a bill in 1940 to authorize it, and Boardman was enthusiastic. However, stockmen were less enthusiastic; and the federal Bureau of the Budget was not enthusiastic at all, fatally undermining its progress.

Owyhee Canyon Scene

that are either threatened or endangered; another thirty-six are vulnerable. But industry has viewed them as hardly different than its lands managed for timber production, and the state has largely acquiesced.

When Neil Goldschmidt was governor, he tried to break domination of the Board of Forestry (which manages the forests) by the timber industry through appointing some environmentally oriented people, such as Janet McLennan and Bob Straub's wife, Pat. But they were still a minority and little change resulted.

Now only twenty percent of the lands in the state forests are earmarked primarily for conservation. It remains to be seen whether a new requirement that a category be set aside for lands of "high conservation values" will result in increasing the amount of lands withdrawn from regular logging.

And state efforts to regulate private timber lands have been weak, emphasizing replanting, with minimal streamside buffers (only a twenty-foot "no touch" zone)—rather than putting more emphasis on protecting fish, wildlife, and habitat. Modest improvements, though, were made in

CHAPTER 11—Conclusions

1987 and 1991, spearheaded by John Kitzhaber, when he led the Senate. Notwithstanding state propaganda, its Forest Practice laws pretend to do more than they accomplish.

In fact, Oregon regulators seem to be inherently timid. They have not distinguished themselves in curbing pollution. They were slow and hesitant in cleaning up the Willamette River and then let pollution in it grow again. In recent years, fifty-nine percent of Oregon's streams have failed to meet federal clean water standards. Pulp mills and other paper processors became serious sources again by the mid-1990s. The Willamette River has been polluted with fecal contamination from Corvallis downstream, and with chemicals below Salem. High levels of mercury have been found in the lower Willamette River in the 1990s, with deformed fish.

In 1993 the Agricultural Water Quality Management Act purported to oblige farmers and ranchers to protect streams abutting their fields from harmful chemicals in runoff, but it has had little effect since farmers must police themselves. Almost half of Oregon's farms are out of compliance or near it. While lottery dollars have restored nearly a quarter of the state's farmland streams, still over 13,000 miles of Oregon's streams fail to meet federal water quality standards. Effluent standards (TMDLs) have not even yet been set for many of Oregon's streams. Fecal contamination levels were so high along many of Oregon's beaches that people were asked to stay out of the water along the central Oregon coast in 2003; NRDC even listed Oregon as one of the two worst states that year (as failing to meet federal standards for safe beaches).

Oregon is below average in inspecting major sources of pollution, according to the EPA. They have set standards for stream temperatures that are too high for salmon, which a federal court has just struck down. In fact, in 2005 the federal EPA even threatened to take over DEQ's wastewater programs because of widespread deficiencies in it.

DEQ let field burning continue for an unconscionably long time.

DEQ also has been unable to get the contaminated Superfund sites in the state cleaned up, despite years of trying. Only four of seventeen have been cleaned up, and eight sites date back to the last (twentieth) century.

Oregon was also found wanting in dealing with the risks associated with incineration of chemical wastes, specifically mustard gas, at the Umatilla Chemical Depot (a U.S. Army facility). These chemicals contained high levels of mercury that were released to the environment when they were burned—not being captured by the filters used. Mercury poses a number of well-known health and environmental risks.

DEQ did not push hard enough to assure observance of Oregon's worthy standard that the "best available alternative" be used—though the state's role was limited. While supposedly a partnership operating under state DEQ permits, the federal role was controlling. For instance, according to the Government Accountability Project, in 2008 it could have called for an analysis of the risks and benefits of each alternative means of disposal. A less troublesome alternative had already been used at the Aberdeen Proving Ground. It should be noted that the Sierra Club, the National Wildlife Federation, and GASP did not ignore this problem. Now at last, this incineration program has ended (it functioned between 2004 and 2011), though the contaminated site has not yet been entirely cleaned up.

In 2011, Oregon's Environmental Quality Commission did make major improvements in the standards governing levels of toxic chemicals permitted in Oregon's waters, which are now the nation's strongest. These are designed to protect those who consume the most fish, particularly tribes. But National Marine Fisheries thinks they are still not strong enough to protect endangered species in Oregon from the following materials: cadmium, aluminum, copper, and ammonia. In addition, DEQ will provide waivers to facilities that cannot meet these standards, such as at industrial sources (subject to EPA approval), and the standards will be difficult to meet on forests and agricultural lands (the legislature has given the agencies that oversee them the lead in enforcement).

With regard to Oregon's air, EPA has found that the ambient levels of air toxins exceed health standards for seventeen of them, with weak enforcement. Its DEQ has never found an industrial plant to be creating a public nuisance, despite frequent public complaints of nauseating emissions from some of them.

Oregon has an inconsistent record in embracing the idea of Integrated Pest Management, used to control pests, on crops and elsewhere. It adopted this as governing policy in 1991, along with a right-to-know law on commercial use in 1999; then dropped it in 2001, and then, in response to OEC's urgings, re-adopted it in 2009. While this policy does require applicators to maintain records (available to the public) and to read pesticide labels, and to provide public notice (except for big applicators, which get general permits), nonetheless it turns on the exercise of judgment in balancing environmental with agricultural goals. It mainly punishes gross negligence and major mishaps. By its nature, it seems to be essentially aspirational, though at one time it epitomized the right approach.

DEQ acts mainly in response to federal mandates, usually waiting until the last moment. To some extent, DEQ's staff often seems to be enfeebled by periodic threats to cut off its funding by hostile legislative committees.

And Oregon never adopted a little NEPA act, as many of the best states did in the early 1970s, though their number is shrinking (from a high of twenty-eight to sixteen now). And many of these are riddled with exemptions (e.g., exempting permitting actions) and allow contractors to prepare EISs. Oregon environmentalists may have not bothered to push for such a measure here because the federal NEPA covered most of the issues that concerned them here (such as proposed freeways and airports, or major timber sales on federal lands). Also, Oregon environmentalists were preoccupied with passing substantive measures (such as the land-use bill).

Oregon's regulatory failures seem to reflect the split of cultures in Oregon between norms of acceptable behavior regarding the environment in rural and urban areas. Environmental norms do not seem to have taken hold in rural areas, or are slow to do so.

No utopias exist in the world in which we live. We learn from our mistakes of omission and commission. But we cannot exult in doing everything right.

Ease of Acceptance

The successes recounted here appear to have become a permanent feature of the Oregon scene. None of them have been repealed or rolled back in their entirety. As time passes, they become part of the way that things are done in Oregon.

Some of the breakthroughs, however, have been accepted more readily than others. Generally the ones that fall in the category of conservation—i.e., dealing with the natural scene (e.g., with parks, wilderness, and wild rivers) or wildlife habitat—seem to be accepted more readily than those that can be categorized as environmental (i.e., dealing with pollution and litter control, energy, and planning).

This may be for a number of reasons. For openers, the idea of conservation and reserves for that purpose are better understood in rural regions. It is not a new or foreign notion. This does not mean that the residents in those areas have agreed with the extent or number of such reserves, but the rationale for them does not seem alien. It has been part of their culture.

But the environmental programs are newer and less understood in rural areas. They strike them as the product of urban regions and are not as much part of their culture.

Moreover, at times some rural residents have seen environmental programs as an attack on their culture or on the power of their local governments. They have fought back at the polls and in the courts. They have not succeeded, but they have waged battles with determination and skill.

Finally, the broader the reach of environmental programs, the greater is the resistance. Some try to reach every corner of the state, or apply to everyday behavior. These are programs such as Oregon's land use planning law and its bottle bill. They have drawn the greatest resistance and most efforts to roll them back. Some see them as changing settled patterns for the distribution of governmental powers, or intruding into areas of private behavior that they think ought to be beyond the reach of government. In that sense, they disturb deeply seated political philosophies.

Those with such philosophies may also feel that programs to curb pollution go too far into adding to the financial burdens imposed on business. As a result, they want to move slowly to be surer of the risk posed by pollutants—waiting for evidence that cannot be denied. They also want to move slowly to allow business to find newer and less expensive ways to cope.

Thus, environmental programs face resistance rooted in rural culture and in conservative political beliefs. Conservation programs do not face such obstacles. The opposition they encounter seems simply to reflect degrees of conflict over access to resources and maintenance of the associated jobs. These are not trivial concerns, but they are much more pragmatic issues, and less rooted in regional culture and politics.

As time passes, these newer additions to the Oregon scene will become part of the accepted culture for all Oregonians. They will seem less foreign and more normal. But then, new additions keep being made to the record, with new patterns of resistance and skepticism. So the "dance" continues.

With regard to conservation issues, Oregon has a longer record of progress, but now it has been moving forward in the environmental arena for over half a century. Neither seems to reflect a passing fad. They have become a permanent part of Oregon's culture.

And most of them have either been initiated by citizen activists or maintained by them. Their activism and determination are equally a part of Oregon's culture.

It is a record in which we can justly take pride.

REFERENCES

Elizabeth and William Orr, *Oregon Water: An Environmental History* (Inkwater Press, Portland, Oregon, 2005).

APPENDIX A
TIMELINE OF CONSERVATION ACCOMPLISHMENTS IN OREGON

DATES	ACCOMPLISHMENTS
1886	Withdrawal of Crater Lake townships
1893	Cascade Range Forest Reserve created
1890s	Cascade Forest Range Reserve defended
1902	Crater Lake NP established
1903–06	Most of withdrawals for Oregon forest reserves
1904–10	Oregon land fraud trials
1907	Three Arches NWR created
	Oregon Caves NM set up
1908	Malheur and Klamath Marshes NWRs established
1913	Oregon beaches reserved for the public
1916	Columbia River scenic highway built
	O & C lands re-vested in federal hands
1921	Gov. Olcott gets legislature to authorize state to preserve scenic beauty along highways by acquiring roadside land
	State law to keep state licensed dams off lower and middle sections of Rogue River
1925	Legislature authorizes state park system to acquire tracts that did not abut highways
1929	Samuel Boardman hired; first director of state parks
1936	Hart Mountain made NWR
1938	Oregon voters set up system to clean up river pollution
1944	Summer Lake WMA set up
1945	Roseburg editor keeps dams off Umpqua River
1946–47	Dams kept off main stem of McKenzie River

APPENDIX A—Timeline of Conservation Accomplishments in Oregon

DATES	ACCOMPLISHMENTS
1947	Forest Park set up in Portland
1950s	Sewage treatment plants built
1955	First mining law reform: Multiple Use Mining Act
1956	Voters defeat EWEB proposal for upper McKenzie dams
1962	High country reserved for recreation under Forest Service High Mountain Policy; logging plans there ended
1964	Finley NWR set up
1965	Baskett Slough & Ankeny NWRs set up
1965–66	Pulp mills forced to stop dumping sulfite wastes
1967	Greenway established along the Willamette River
1968	Lower Rogue River made a wild river; among first set up
1969	State Supreme Court upholds public's right to cross dry beach sands
	State drops plans to build series of freeways along coast; outgrowth of battles over Nestucca Spit highway
1969–71	Bottle bill enacted requiring deposit
	Eugene voters turn down EWEB nuclear plants
1970	Initiative passed protecting state scenic waterways
	Wigwam burners phased out under new air pollution laws
1970s	Over-grazing curbed in Malheur NWR
1972	Oregon Dunes NRA established
	South Slough Estuarine Reserve set up
	Minam Valley added to Wallowa wilderness
1973	State land use legislation (SB 100) enacted
	Expansion of Portland Airport blocked
1974	Portland's proposed Mt. Hood freeway defeated
	Congress sets up Cascade Head Scenic Research Area

DATES	ACCOMPLISHMENTS
1975	Hells Canyon NRA established
	John Day Fossil Beds NM established
1978	387,000 acres added to Oregon wilderness, including 45,500 acres at French Pete Creek
1980	State voters pass initiative measure posing impediments to any more nuclear plants
	Northwest Power Act enacted by Congress
	Congress improves boundaries for Crater Lake NP
1982	Congress buys out Rock Mesa mining claims
1983	Frontage along lower Deschutes River acquired
	Spraying of dangerous herbicides in coastal national forests stopped
1984	861,500 acres added to Oregon wilderness areas in 31 units, including first oriented toward old growth habitat
1986	Columbia River Gorge Scenic Area established
1987	Oregon Endangered Species Act passed
1988	40 stream segments made wild rivers under Oregon Omnibus Wild & Scenic Rivers Act
1989	Law passed to protect state's groundwater
1990	Newberry National Volcanic Monument established
1991	Grazing suspended in Hart Mountain NWR
	State regulates surface mining using chemical processing
1991–2009	Field burning largely phased out
1992	Eugene wetlands saved
	Tualatin River NWR authorized
1993	Clinton proclaims Northwest Forest Plan
	Portland's Heritage Tree Program started

Appendix A—Timeline of Conservation Accomplishments in Oregon

DATES	ACCOMPLISHMENTS
1996	Trojan nuclear plant closed
	Congress ends logging in Bull Run watershed
	Opal Creek wilderness set aside
2000	Steens Mountain Cooperative Management and Protection Area established
	Cascade-Siskiyou National Monument established
	Zumwalt Prairie Natural Area established
2006	Senator Wyden causes benzene levels in air to be phased down
2007	Standards set for renewable power (25% by 2025)
	Legislation requires recycling of discarded electronics
	Disposable water bottles put under Oregon bottle bill
2008	Judge James Redden orders release of more water for salmon from dams
	Elk Creek dam on the upper Rogue breached
2009	Standards set that rule out new coal-fired power plants
	223,400 acres added to wilderness system in 12 units, especially around Mt. Hood
	35,000 acres set aside as NRA at Mt. Hood
	10% reduction required in carbon content of vehicle fuels
	30% less greenhouse gas emission for vehicles by 2016
	Destination resorts ruled out of Metolius basin
2011	Portland completes its "Big Pipes"
	More beverage containers put under Oregon's bottle bill
	First two state marine reserves established
2012	Three more state marine reserves established

APPENDIX B
MAP OF PLACES MENTIONED IN THE TEXT

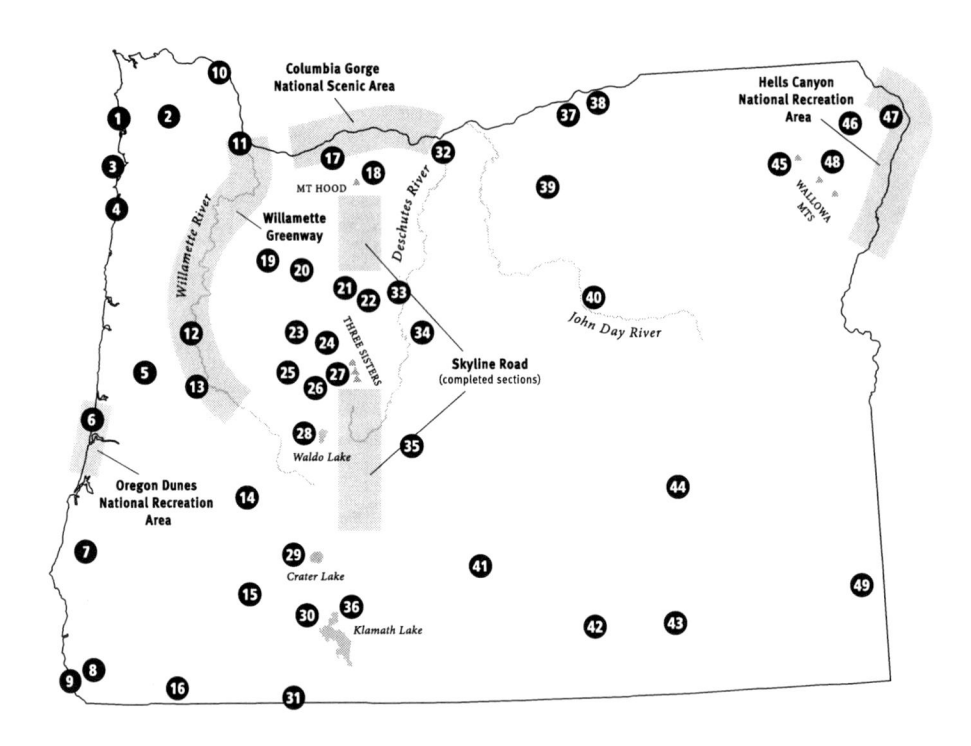

1. Place where Os West Resolved to Protect Public Access to Oregon's Beaches
2. Tillamook and Clatsop State Forests
3. Three Arch Rocks National Wildlife Refuge
4. Nestucca Spit
5. Site of Worst Herbicide Spraying
6. Oregon Dunes National Recreation Area
7. O & C Lands
8. Siskiyou region; also Kalmiopsis Wilderness

Appendix B—Map of Places Mentioned in the Text

9. Curry County coast
10. Trojan Nuclear Plant (now removed)
11. Forest Park (Portland)
12. Field Burning
13. West Eugene Wetlands
14. Umpqua River
15. Elk Creek Dam
16. Oregon Caves National Monument
17. Bull Run basin
18. Mt. Hood
19. Silver Falls State Park
20. Opal Creek Wilderness
21. Mt. Jefferson Wilderness
22. Metolius Basin
23. South Santiam River
24. Site of Defeated EWEB Project on Upper McKenzie River
25. Main Stem of the McKenzie River
26. French Pete Creek and the Deleted 53,000 Acres
27. Three Sisters Wilderness
28. Waldo Lake
29. Crater Lake National Park
30. Sky Lakes Wilderness
31. Cascade-Siskiyou National Monument
32. Lower Deschutes River
33. Round Butte Dam
34. Smith Rock State Park
35. Newberry National Volcanic Monument
36. Upper Klamath Marsh National Wildlife Refuge
37. Boardman Coal-Fired Power Plant
38. Umatilla Chemical Depot

39. Shepherd's Flat Wind Farm
40. John Day Fossil Beds National Monument (Visitor Center)
41. Summer Lake State Wildlife Management Area
42. Hart Mountain National Wildlife Refuge
43. Steens Mountain Complex
44. Malheur National Wildlife Refuge
45. Minam River
46. Zumwalt Prairie Natural Area
47. Hells Canyon National Recreation Area
48. Wallowa Wilderness
49. Owyhee Canyons

APPENDIX C
LIST OF ORGANIZATIONS THAT MADE CONSERVATION HISTORY IN OREGON

The Mazamas: establishment of Crater Lake National Park

Portland Audubon Society: early federal waterfowl refuges in Oregon and others set up at later time in Willamette Valley

Oregon Duck Hunters Association: establishment of Willamette Valley wildlife refuges

National Wildlife Federation, Oregon chapter: initiative measure to curb water pollution in 1930s; defense of McKenzie River

Oregon Roadside Council: growth of state parks

Federation of Western Outdoor Clubs (FWOC), including Mazamas, Trails Club: Portland's Forest Park; and Mt. Jefferson Wilderness (Chemeketans)

Izaak Walton League, Oregon chapter: defeating dams and cleaning up pulp plant discharges; advances of multiple use on O & C lands

Oregon Environmental Council: many pioneering environmental measures, including bottle bill, land use legislation, surface mining law

1000 Friends of Oregon: defense of Oregon's land use law

Hells Canyon Preservation Society: establishment of the Hells Canyon National Recreation Area

Sierra Club, Oregon chapter: French Pete Creek wilderness and many wilderness gains, including Steens Mountain complex

Oregon Wildlife Heritage Foundation: protecting banks of the lower Deschutes River

Oregon Natural Resources Council/Oregon Wild: various wilderness gains, wild river reservations, and notching Elk Creek dam

Lane County Audubon Society: early efforts to protect Spotted Owl and old growth habitat on BLM lands

Corvallis Audubon Society: restoring habitat in the Malheur refuge

Citizens Against Toxic Sprays (CATS): ending aerial herbicide spraying of federal forests

Friends of the Columbia Gorge: Columbia Gorge Scenic Area designation

The Nature Conservancy, Oregon chapter: preserving various natural areas, including the Zumwalt Prairie

Native Plant Society, Oregon chapter: state program to protect endangered species

Northwest Environmental Advocates: Portland's big pipes

Oregon League of Conservation Voters: 2007 Electronics Recycling Program

Oregon Natural Desert Association: Oregon Badlands Wilderness and Spring Basin Wilderness (BLM)

Our Ocean Coalition: state Marine Reserves

Ad Hoc Groups That Were Organized Around a Campaign

McKenzie River Protective and Development Association: defeat of main stem dam on the McKenzie River

Save the McKenzie Association: defeat of a dam on the upper McKenzie

Friends of the Three Sisters Wilderness: defense of the Three Sisters Wilderness

Appendix C—List of Organizations That Made Conservation History in Oregon

Citizens Committee to Save Our Beaches: keeping freeways off coastal beaches

Citizens Committee to Save Our Sands: defense of Nestucca Spit

Bull Run Interest Group (BRIG): end to logging in Bull Run watershed

Newberry Crater Stakeholders Discussion Group: establishment of the Newberry National Volcanic Monument

INDEX

Symbols

2,4,5-T 129-131
1000 Friends of Oregon 42, 91
1855 Pacific Railway Survey 149

A

Aberdeen Proving Ground 217
ACECs. *See* Areas of Critical Environmental Concern
Adams, Brock 147
Aerosol sprays 100
AFL-CIO 88, 105
Agent Orange 129, 131
Agricultural Water Quality Management Act 216
Agriculture Department 31, 47, 170
Ahwanhee Lodge 48
Air Pollution Authority 98
Air quality 98-100, 119
Alsea 131
Alvord Desert 152, 154
American Fisheries Society 202
American Forestry Association 6
American Lung Association 100
Anderson, Jack 130
Anderson, Jean 129
Ankeny Reserve 162, 165, 223
Antelope 31-33, 35
Antiquities Act 154, 155
AOI. *See* Association of Oregon Industries
Applegate River 57, 132
Arch Cape 39
Areas of Critical Environmental Concern 159, 160
Areas of Critical State Concern 93, 94

Ashland 188
Association of Oregon Counties 105
Association of Oregon Industries 86, 88
Atiyeh, George 186
Atiyeh, Vic 97, 186-188
Atkinson, Jason 105
Atomic Energy Commission 113
AuCoin, Les 68, 150, 174, 184, 212
Audubon Society
 Corvallis 166, 230, 235
 Lane County 124, 182-184, 196, 230, 239
 Oregon 27, 165, 241
 Portland 27, 163, 165, 172, 198, 201, 206, 229, 243
 Seattle 197, 206
Aurora 120
Avakian, Brad 105
Axline, Michael 196

B

Babbitt, Bruce 153, 155, 156
Bacon, Robert 81, 82
Badger Creek 184, 190
Bagby Hot Springs 176
Baker City 7, 12
Bald Eagles 30, 159
Bandon Marsh 25
Barkan Dunes 138
Barred Owl 200
Baskett Butte 162
Baskett Slough Refuge 162-165, 223
Beaches ii, 36, 37, 50, 80, 81, 82, 83, 84, 112, 121, 122, 123, 208, 211, 216, 222, 226, 230

232 | Conserving Oregon's Environment

Index

Public 80
Wet 80
Beaches Forever 50, 81
Beacon Rock 146
Bear 151
  Black 157
Bear Valley National Wildlife Refuge 30
Beaver 127
Beck, Borden 120
Beck, Lucille 44
Bend 13, 34, 38, 64, 149, 168, 188-190
Bennett, E. H. 43
Benson High School 122
Benson, Simon 60
Benton County 128, 165
Benzene 101, 102, 225
Bergel, Peter 116
Berger, Vicki 86
Biddle's Lupine 152
Big Creek 115
Big Obsidian Flow 151
Biological Survey 28, 31, 33
Birds 26-28, 30, 33, 107, 110, 151, 155, 157, 163, 166-168, 201
Bitter Fog 129
Black Butte 94
Blackstone 83
Blitzen watershed 154
Blitz-Weinhard brewery 85
BLM. *See* Bureau of Land Management
Blomquist, Jim 193
Bluebunch Wheat Grass 157
Blue-Gray Gnatcatcher 155
Blue Mountain Reserve 7, 11
Blue Mountains 7-12
Blue River 55
Blumenauer, Earl 189
Boardman coal-fired power plant 107-109, 227
Boardman, Sam 38-40, 42, 145, 213, 214, 222
Board of Forestry 215
Bobcats 190
Bohlman, Herman 26, 28
Bonine, John 196
Bonneville Dam 204

Bonneville Power Administration 107, 192-194, 202, 204
Booth, Brian 40
Booth-Kelly Company 45
Boring 115
Bottle bill 87, 103, 210, 219, 223, 225, 229
Boulder Creek 177, 180, 181, 185
Boy Scouts 43, 44
BPA. *See* Bonneville Power Administration
BRIG. *See* Bull Run Interest Group
Brookings 214
Brower, David 70
Bull-of-the-Woods 185, 190
Bull Run 2, 173-175, 225, 227, 231
Bull Run Interest Group 174, 231
Bull Run Trespass Act 174
Bull Run watershed 2, 225, 231
Bureau of Forestry 8, 10
Bureau of Land Management 20, 25, 57, 68, 98, 121, 122, 127, 131, 132, 153-156, 159, 188-190, 196, 199, 230
Bureau of Public Roads 48
Bureau of Reclamation 28
Burns 152
Bush, President George H. W. 101
Bush, President George W. 200
Butte Creek 57
Butterflies 155, 157
Buxton family 121

C

Cabell, Henry Failing 164
California Energy Company 150
Camp Fire Girls 44
Canada Goose 162, 164
Canby 95
Cannon Beach 36, 37, 80, 81
Cannon, Garnett "Ding" 43, 44
Cant Ranch Museum 146
Canyon City 7, 11
Canyon Wren 155
Cape Falcon 158
Cape Kiwanda 115, 121
Cape Lookout Park 39
Cape Meares 25

Michael McCloskey | 233

Cape Perpetua 139, 158
Carmen-Smith project 55
Carnegie Institution 47
Carter, President Jimmy 119, 180, 193
Cascade Development Company 47
Cascade Head 158, 159, 223
Cascade Head Catchfly 159
Cascade Head Experimental Forest 159
Cascade Head Scenic Research Area 223
Cascade Locks 62
Cascade (Range Forest) Reserve 1, 5, 13, 14, 16, 34, 222
Cascades xiii, 1, 3, 4, 16, 28, 34, 48, 54, 64, 67, 72, 74, 77, 79, 94, 155, 182, 214
  Central 94
  Central Oregon 54, 64, 67, 74, 214
  North 67, 72, 77
Cascade-Siskiyou National Monument 154, 188, 190, 209, 225, 227
Cascades Recreation Area 182
Cascadia Dam 57
CATS. *See* Citizens Against Toxic Sprays
Cattle 1, 7, 8, 13, 96, 98, 153, 154, 156, 166
Cattle grazing 153, 154, 156, 166
Cave Junction 159
CCC. *See* Civilian Conservation Corps
Celilo Falls 213
Cenozoic Era 144
Central Oregon Cascades 54, 64, 67, 74, 214
Century Drive 48
Chamberlain-Ferris Act 22, 23
Chamberlain, George 8, 9, 15, 22
Chambers, Richard 83, 84, 86
Charbonneau 89
Charlton, Dave 52
Checkerboard lands 20, 45
Cheetah 31
Chemeketans 66, 229
Chemicals 110, 113, 216, 217
Chief Joseph Dam 203
Church, Frank 143
Citizens Against Litter 85
Citizens Against Toxic Sprays 129, 131, 230

Citizens Committee for the Columbia River 119
Citizens Committee to Save Our Beaches 230
Citizens Committee to Save Our Sands 121, 231
Citizens Utility Board 105
City Club of Portland 43, 51
City Council, Portland's 118
Civilian Conservation Corps 25, 42, 139, 161
Clackamas River 176
Clackamas Wilderness 190
Clark, Don 118
Clarno 146
Clatsop State Forest 214, 226
Clean Air Act 99, 118
Clean Cars Program 110
Clean Energy Coalition 105
Clean Water Act 135
Clear-cuts 175
Cleator, Fred 43, 48
Cleveland, President Grover 1, 4-6, 17
Clinton, President Bill 153-155, 198-200, 224
Cloud Cap Inn 10
Coal projects 107
Coastal sphagnum moss 25
Coast Range 14, 159, 201
Coburn, Tom 189
Code of Federal Regulations 68
Columbia basin 202
Columbia Gorge 59-62, 93, 106-108, 146, 147, 208, 209, 230
Columbia Gorge Highway 62
Columbia Gorge United 147
Columbia River 2, 36, 48, 59-63, 93, 94, 106-108, 113, 119, 135, 136, 146-148, 175, 185, 193, 194, 202-204, 208, 209, 222, 224, 230
Columbia River Gorge 59-62, 93, 106-108, 146, 147, 152, 208, 209, 224, 230
Columbia River Inter-Tribal Fish Commission 194
Columbia Slough 135, 136
Columbia system 203
Combined-sewer-overflow outfalls 136

Committee to Save the Beaches 122
Common Murres 26
Condon, Thomas 145, 146
Congress 1, 3, 5, 6, 18-23, 30, 47, 57, 68, 74-76, 98, 101, 132, 144, 147, 149, 150, 154, 159, 167, 174, 175, 183, 186, 193, 198-200, 204, 223-225
Conservation Commission 15
Consumer Affairs Committee 86
Cook, Stan 124
Cooper Spur 190
Coos Bay 21, 138, 141, 159
Coos-Bay Wagon Road lands 23
Copper River Delta 162
Copper-Salmon (Wilderness) 188-190
Cornucopia 11
Corps of Engineers 53, 55, 57, 134, 204
Corvallis 34, 35, 50, 87, 127, 128, 162-166, 191, 206, 216, 230
Corvallis Audubon Society 166, 230
Cottage Grove 94
Cougar 157
Cougar Creek 55
Cougar Lakes 72
Council Oak 171
Cove Palisades State Park 213
Cows 12, 33, 154
Cox, Thomas 39
Crabbing 158
Crater Lake xi, xii, 16, 17, 19, 34, 48, 149, 181, 208, 222, 224, 227, 229
Crater Lake National Park xi, 16, 17, 34, 48, 149, 181, 208, 227, 229
Crater Lake Reserve 19
Crescent Lake 48
Crooked River 13, 42, 213
CSO. *See* Combined-sewer-overflow outfalls
Curry County 214, 227
Cutler, Rupert 130, 180
Cyanide 110

D

Dams 28, 30, 53-60, 62, 107, 132, 138, 141-143, 157, 175, 193, 194, 202-205, 210-213, 222, 223, 225, 229

Dana, Marshall 32
Darling, "Ding" 33
Davis 20
Day, L. B. 90
Days Creek 57
Dayville 146
Deer 190
DeFazio, Peter 189, 212
Dellenback, John 141
Department of Environmental Quality 90, 98, 101-103, 107, 109, 135, 136, 216-218
Department of Fish and Wildlife 168, 170, 196, 203, 210
Depression 32, 42, 48, 49, 51
DEQ. *See* Department of Environmental Quality
Deschutes National Forest 2, 94, 149
Deschutes River 96, 97, 148, 213, 224, 227, 229
Deschutes River canyon 96
Detroit Lake 185
Diamond Crater 159
Diamond Peak 66, 77, 182, 185
Dioxin 129, 130
Division of State Lands 59, 159, 170
Douglas County 181
Douglas Fir 160
Douglas, William O. 54, 70, 141
Dragonflies 127
Drake, June 40
Drift Creek 184
Dr. Pepper 85
Ducks 28, 31, 107, 167
  Wood 167
Duck Stamp Act 168
Duncan, Robert 140, 174
Dunes 138, 139, 141
Durning, Marvin 120
Dusky Canada Goose 164
Dutton, Clarence 16-18
Dwyer, Judge William 198, 199, 236

E

Eagle Cap Wilderness 69, 71, 77, 78
Eagles 30, 31, 151, 159

Early Blue Violet 159
Earth Day 207
Eastbank Esplanade, Portland's 101
Eastern Oregon College 70
Eastern Oregon Land Company 97
East Lake 149
East Side Company 20
Eber, Ronald ix, 34, 141
EIS. *See* Environmental Impact Statement
Eisenhower, President Dwight D. 14, 173
Elijah Bristow State Park 95
Elk 158, 190
Elk Creek 132-134, 159, 225, 227, 230
Elk Creek Dam 132, 133, 225, 227, 230
Elk River 190
Elliott State Forest 46
Endangered species 98, 127, 134, 135, 138, 155, 158, 162, 170, 175, 182, 195, 197, 202-204, 215, 217, 224, 230
   State efforts to protect 170
Endangered Species Act 134, 135, 170, 195, 197, 198, 203, 224
Energy Issues 104
Enterprise 144
Environmental impact statement 118-120, 129, 130, 183, 196
Environmental Quality Commission 100, 101, 217
EPA 100-102, 107, 126, 130, 131, 216, 217
ESA. *See* Endangered Species Act
Eugene 34, 52, 55, 64, 66, 71, 79, 84, 87, 95, 100, 105, 115, 124-131, 140, 177, 178, 196, 209, 223, 224, 227
Eugene Water and Electric Board 55, 105, 115, 223, 227
European beach grass 138
Evans, Brock ix, 65, 141, 142, 179
E-waste 102, 103
EWEB. *See* Eugene Water and Electric Board

F

Fadeley, Nancie 90
Failing Estate 164
Falcon 155

Farmers Political Action Committee 90
Federal Coastal Zone Management Act 93
Federal mining laws, reform in 69
Federal Power Commission 141
Federal Reserves, Foundational 1
Federal Waterfowl Refuges 162
   Willamette Valley 162
Federation of Western Outdoor Clubs 43, 66, 74, 76, 140, 229, 248
Fender's Blue Butterfly 127, 162
Ferguson, Denzel 166
Field burning 99, 224, 227
Finley refuge 162-165
Finley, William 26-28, 30, 35, 163, 223
Fish 57, 94, 114, 119, 132, 134, 136, 143, 152, 155, 182, 192, 193, 202-205, 209, 215, 216, 217
   Deformed 216
   Hatchery 202
Fish and Wildlife Act 168
Fish and Wildlife Service 28, 31, 33, 134, 163, 195, 197, 198, 200
Fishing 57, 59, 60, 94, 97, 158, 193, 213
   Drag 158
Floatage easement 59
Florence 115, 140
Folts, Mert 54
Forest Park 43, 44, 50, 209, 223, 227, 229
Forestry Commission 6, 19
Forest Service 16, 23, 24, 34, 43, 44, 47, 48, 50, 57, 61, 62, 64-76, 94, 129-132, 139-143, 147-150, 155, 161, 173, 174, 176, 178, 180, 182, 183, 186, 191, 195-198, 206, 209, 223
Forsman, Eric 195
Fossil 146
Fossils 144-146, 160
FPC. *See* Federal Power Commission
Franklin, Jerry 201
Fred Meyer 87
Fremont National Forest 14
Fremont-Winema National Forest 14
French Pete Creek 67, 73, 176-182, 224, 227, 229
Frenkel, Bob and Liz 128
Friends of Forest Park 44

INDEX

Friends of the Columbia Gorge 108, 146, 230
Friends of the Earth 146
Friends of the Three Sisters Wilderness 66, 68, 230
Friends of the [Tualatin River] Refuge 167
Frogs 127
Fulton, Charles 9, 15
FWOC. *See* Federation of Western Outdoor Clubs

G

Gabrielson, Ira 31-33, 35
Game Commission 31, 32, 165
GAO. *See* General Accounting Office
Garrett, Stuart 149-151
GASP 217
Gearhart Mountain 77
Geese 107, 162, 164, 165, 167, 168. *See also* Goose
General Accounting Office 57, 165
General Land Office 2, 4-6, 8, 9, 12, 14, 23, 24, 45-47
General Mining Law of 1872 69
Gentner's Fritillary 155
Geological Survey 8, 10, 16, 18
Gilliam County 13, 106
Glass, Gordon 144
Glazer, Jane 171
Glen Avon bridge 98
Godfrey, Arthur 142
Gold 110
Golden Paintbrush 162
Gold Hill Diversion 132
Gold Ray 132
Goldschmidt, Neil 40, 118, 119, 212, 215
Goodwin, Alfred T. 82
Goose
    Semidi Islands Aleutian Cackling 25
Goose Lake 13
Gordon, Steve 126, 127
Gore, Vice President Al 198
Gorton, Slade 147
Government Camp 190
Grand Canyon of Oregon 213
Grand Coulee Dam 203
Grande Ronde River 12
Grant County 7, 144
Grants Pass 105, 177
Grasses 157
Grassy Knob Wilderness 190
Gray, John 88, 91
Great Basin 152, 155, 157
Great Basin Wild Rye 157
Great Egrets 28
Greater Sage Grouse 33
Greater Sandhill Cranes 30
Great Grey Owl 155
Greeley, William B. 47, 66
Greene's Mariposa Lily 155
Greenway program 93
Greenways 93, 94, 96, 223
Grey, Zane 57
Griffith, Emerson 48
Groundwater 102-104, 210, 224
Group of Fifty 44
Grouse 33, 155

H

Habitat xiii, 1, 25-28, 30, 31, 33, 39, 54, 59, 111, 119, 121, 124, 127, 128, 132, 138, 143, 155, 159, 160, 162, 163, 168, 169, 175, 190, 191, 194, 196, 198, 200-202, 204, 209, 211, 213, 215, 219, 224, 230
Habitat Conservation Areas 198
Hall, John Hicklin 46, 47
Hallock, Ted 89, 90
Halprin, Lawrence 88
Hampson, Al 84, 85
Hanneman, Paul 84
Hansell, Stafford 59
Harbor Drive 118
Harriman, E. H. (Ned) 31, 46
Harriman Springs 31
Hart Mountain 31-33, 222, 224, 228
Hart Mountain Antelope Range 33
Hart Mountain Antelope Refuge 31
Hatfield, Mark 19, 58, 68-71, 76, 132, 140-143, 147, 149, 151, 153, 165, 167, 174-178, 180-188, 190, 193, 212, 213

Michael McCloskey | 237

Hayakawa, S. I. 184
Hay, William 80-83
Hells Canyon 141-143, 156, 157, 203, 214, 224, 228, 229
Hells Canyon National Recreation Area 141, 142, 228, 229
Heney, Francis 20, 46
Heppner Reserve 12, 13
Herbert, Sydney ix, 124, 183, 184, 196
Herbicides 128-131, 224
Herbicide spraying 128, 226
Heritage Tree Program 248
  Portland's 171, 224
Hermann, Binger 8, 9, 14, 17, 46
Hermit Warbler 155
Hewitt, James P. 186
High Mountain policy 73, 223
Highway 101 39, 123, 138, 139
Highway Commission 38, 117, 121, 122, 123
Highway Department 39, 40, 117, 120
Highway Trust Fund 117
Hill, Bonnie 131
Hill, Sam 60
Hitchcock, Ethan 46
Hogan, Michael 30
Honeyman, Jesse M. 38, 39
Honeyman State Park 139, 141
Hopson, Ruth 66
Horse Creek 65, 66
HUD 89
Hyde, Phil 70
Hydro 58, 60, 97, 104, 106, 192
Hydro-power 192

I

Ickes, Harold 24
Idaho 12, 141, 143, 157, 193, 203
Idaho Fescue 157
Idaho Power Company dam 203
Imnaha Reserve 12
Important Bird Areas 33
Indian fishing sites 213
Indian reservations 14, 15, 30, 74
Indian tribes 44, 147, 148, 193, 217

Industrial Customers of Northwest Utilities 105
Integrated Pest Management 218
Interior Board of Land Appeals 68
Interior Department 2, 20, 22-25, 47, 196
Interior Secretary 7, 30, 69, 122, 153
Intermediate Recreation Area 178
Izaak Walton League 25, 51-55, 132, 229

J

Jackson-Frazier wetland 127
Jackson, Glenn 88, 91, 121, 122
Jackson, Henry 71, 72
Jaguars 160
Jawbone Flat 186
Jefferson Park 75
Jensen, Gertrude Glutsch 62
John Day Fossil Beds 144, 145, 224, 228
John Day Fossil Beds National Monument 144, 228
John Day River 12, 181, 190
Johnson, Charles 116
Johnson, Lee 85, 86
Johnson, Robert Underwood 6
Jolley, Russ 146
Jones, Holway 178, 182
Joseph 71
Justice Department 20

K

Kalmiopsis Wilderness 77, 177, 180, 181, 226
Kent, W. H. B. 12
Kerr, Andy ix, 63, 79, 153, 183, 185, 191
Kiger Gorge 152
Kincaid's Lupine 162
King City 167
Kitzhaber, John 109, 158, 212, 216
Klamath County 13, 30
Klamath Indian Reservation 14, 30
Klamath Marsh 28, 222
Klamath Refuge 28, 30, 31
Koosah Falls 55, 56
Kopetski, Mike 187
Kulongoski, Ted 94, 105, 124, 158, 212

L

Ladd, W. M. 10
Lafferty, A. W. 20, 21, 22, 46
La Follette, Cameron 183
La Grande 70
Lake County 13
Lakeview 13, 32, 35, 168
Lancaster, Sam 60
Land and Water Conservation Fund 95
Land Conservation and Development Commission 89, 96
Land Frauds 44
Land use 87-90, 93, 94, 128, 148, 210, 212, 219, 223, 229
Land use law 87
Lane Council of Governments 126
Lane County 124, 182-184, 196, 230
Lane County Audubon Society 124, 182-184, 196, 230
Languille, H. D. 8-14
Laurelhurst Freeway 117
LCDC. *See* Land Conservation and Development Commission
Leaburg 55
League of Oregon Cities 90, 105
League of Women Voters 88, 89, 98
Le Conte, Joseph 16
Lewiston, Idaho 12
Lilly Lake 141
Lincoln City 158
Linn County 89
Litter 83-87, 210, 219
  Citizens Against 85
Little Sandy River 175
Logging roads 64, 174, 175, 178, 201
  Conflicts over 175
Lost Creek dam 132, 134
Lost Lake 48
Low Carbon Fuel Standard 109
Lower White River Wilderness 190
Lubchenco, Jane 204

M

Macpherson, Hector 89, 90
Malheur County 103
Malheur Lake Refuge 28
Malheur Marsh 28, 222
Malheur National Forest 11
Malheur (National Wildlife) Refuge 28, 164, 223, 228, 230
  Disputes over grazing in 166
Marbet, Lloyd 115
Marbled Murrelet 201
Marine Reserves 158, 210, 225, 230
Marion County 92
Marion Lake 76
Mark O. Hatfield Wilderness 190
Marshall, David 163-167, 172
Marshall, Robert 64, 65
Martin, Charles 51, 211
Maser, Chris 201
Materials Act of 1947 69
Maury Mountain Reserve 13
Mazamas 4, 5, 16, 43, 44, 66, 189, 229
McArthur, Lewis A. 48
McBride, George 18
McCall, Tom 53, 71, 80, 82, 84, 86, 87, 89-91, 94-96, 99, 100, 112, 118, 119, 122, 123, 211
McFadden's Marsh 164
McHarg, Ian 87, 89
McKay, Douglas 69
McKenzie River 53-56, 222, 223, 227, 229, 230
McLennan, Janet 81, 82, 215
McNary, Charles 23, 33, 214
McRae bill 5, 6
Meier, Julius 60
Mercury 102, 108, 216, 217
Merkley, Jeff 105, 212
Merriam, C. Hart 19
Merriam, John C. 47, 145
Metolius Basin 93, 94, 212, 225, 227
Metolius Conservation Area 94
Metolius River 93
Metro 170
Middle Santiam 180, 181, 185
Migratory Bird (Hunting and Conservation Stamp) Act 168
Migratory Bird Treaty 168
Miller and Lux 7
Miller, Joaquin 160
Miller, Joseph 174

Minam River 228
Minam River Valley 69, 70, 223
Mining laws 69
Minto, John 5, 6
Miscarriages 129-131
Mische, Emmanuel 43
Mitchell 145
Mitchell, John 11, 21, 46, 47
Model State Bird Law 27
Modoc forest reserve 13
Molalla River 95, 98
Molalla River State Park 95
Montreal Protocol 101
Monument
  Cascade-Siskiyou National 154, 188, 190, 209, 225, 227, 234
  John Day Fossil Beds National 144, 228, 238
  Newberry National Volcanic 149, 209, 224, 227, 231, 241
  Oregon Caves National 160, 227, 242
Monument (Oregon) 12
Morrow County 13, 106
Morse, Wayne 57, 66, 67, 72, 76, 82, 139, 140, 212
Moses, Robert 43, 117
Mountain Lakes 77
Mt. Bachelor 48
Mt. Hood 2, 4, 10, 47, 48, 62, 73, 77, 117, 118, 120, 173, 177, 181, 184, 188-190, 214, 223, 225, 227
Mt. Hood National Forest 2, 190
Mt. Hood Wilderness 190
Mt. Jefferson 74, 75, 77, 185, 227, 229
Mt. Jefferson Wilderness 74, 227, 229
Mt. Rainier National Park 19, 45
Mt. Thielsen 182, 185
Mt. Washington 66, 77, 185
Muddy Creek 164, 165
Muir, John 4-6, 16, 19, 31, 34, 59
Multiple Use Mining Act 69, 223
Multnomah County 60, 62
Multnomah Falls 146
Munger, Thornton 44, 50
Myers, Clay 88

N

National Academy of Science 6, 19
National Conservation Area 153
National Environmental Policy Act 118, 131, 195, 218
National Estuarine Areas 159
National Forest Management Act 195, 198, 199, 201
National Forestry Commission 6
National Forests 1, 2, 11, 64, 79, 173, 191
  Origins of the 1
National Historic Landmark 48
National Marine Fisheries 134, 217
National Marine Fisheries Service 134
National Natural Landmarks 149
National Oceanic and Atmospheric Administration 159, 202-205, 210
National Park Service 19, 40, 42, 61, 63, 64, 138, 139, 144-146, 149, 161, 214
National Recreation Area 138, 140-142, 144, 190, 226, 228, 229
National Seashore 138, 139
National Wilderness Preservation System 191
National Wildlife Federation 54, 105, 197, 203, 217, 229
National Wildlife Refuges 25
National Wildlife Refuge System Administration Act 168
National Wildlife Refuge System Improvement Act 168
Native Plant Society 170, 230
Natural Area 128, 156, 225, 228
Natural Desert Association 153, 188, 189, 230
Natural Heritage Bank 128
Natural Heritage Conservation Areas 210
Nature Conservancy 93, 124, 126, 127, 156, 158, 230
  Oregon 93
Nehalem 36, 158
NEPA. *See* National Environmental Policy Act
Neskowin 25, 84, 159
Neskowin Marsh 25

Index

Nestucca Bay 25
Nestucca River 121
Nestucca Spit ix, 50, 120, 121, 123, 223, 226, 231
Neuberger, Maurine 140
Neuberger, Richard 14, 55, 66, 67, 69, 139, 140, 211
Newberry Crater 149, 150, 224, 231
Newberry Crater Stakeholders Discussion Group 231
Newberry, John S. 149
Newberry National Volcanic Monument 149, 209, 224, 227, 231, 241
New Deal 42, 51, 62
Newport 36, 158
New Yorker 130
New York Natural History Museum 64
NFMA. *See* National Forest Management Act
NOAA. *See* National Oceanic and Atmospheric Administration
No Net Loss 125
North Cascades Conservation Council 67
Northern Pacific 45
Northern Spotted Owl 155, 194, 195, 197, 198, 200, 206
North Fork of the Elk River 190
North Fork of the John Day 12, 181, 184
Northwest Environmental Advocates 135, 230
Northwest Forest Plan 199-201, 209, 224
Northwest planning council 193
Northwest Power Act 192, 224
Noyes, Richard 178
NPS. *See* National Park Service
NRDC 105, 216
Nuclear power plants 113

O

Obama, President Barack 200, 204
Obsidian Flow, Big 151
Obsidians 66
O & C Company 20-25, 35, 45, 46, 199, 200, 222, 226, 229
Ochoco National Forest 13
OEC. *See* Oregon Environmental Council

Off-road vehicles 153, 154, 156
Olcott, Ben 37, 38, 211, 222
Old growth 31, 39, 151, 155, 159, 176, 182, 183, 185, 186, 194-201, 210, 212, 213, 224, 230
Olmsted, John C. 43
Olmsted, Jr., Frederick Law 47
OMARK Industries 88
Omnibus Public Land Management Act 189
ONRC. *See* Oregon Natural Resources Council
Onthank, Karl 55, 66, 67, 71, 96
Onthank, Ruth 66
Opal Creek 76, 185-188, 201, 225, 227
Opal Creek Wilderness 185, 227
Oregon Alpine Club 4, 16
Oregon and California Railroad Company 20. *See also* O & C Company
Oregon Anti-Stream Pollution League 51
Oregon Audubon Society 27, 165
Oregon Badlands 188, 190, 230
Oregon Business Alliance 105
Oregon Cascades Conservation Council 67
Oregon Cascades National Park
    Failed campaign for xiii
Oregon Caves 159-161, 222, 227
Oregon Caves National Monument 160, 227
Oregon Central Railroad Co. 20
Oregon City 52
Oregon coast 25, 39, 115, 120, 159, 188, 214, 216
Oregon Coast Association 123
Oregon Conservation Network 105, 158
Oregon Duck Hunters Association 163, 165, 229
Oregon Dunes 138, 139, 223, 226
Oregon Dunes National Recreation Area 138, 226
Oregon Electronics Recycling 102
Oregon Environmental Council 19, 57, 76, 84, 85, 88-91, 93, 102, 105, 108-110, 113, 114, 119-121, 129, 142, 174, 176, 177, 218, 229
Oregon Fish and Wildlife Commission 27
Oregon Game Commission 31

Michael McCloskey | 241

Oregon Highway Users Conference 82
Oregonian 7-10, 14, 19, 52, 117, 142, 176, 206
Oregon Islands 25
Oregon Journal 7, 32
Oregon Land Fraud Trials 46, 50
Oregon League of Conservation Voters 102
Oregon Municipal Utilities Association 105
Oregon Natural Desert Association 188, 189, 230
Oregon Natural Resources Council 58, 63, 79, 153, 182-184, 187, 191, 194, 196, 197, 230
Oregon Nature Conservancy 93
Oregon Omnibus Wilderness Act 176, 191
Oregon Parks and Recreation Commission 40
Oregon Parks Department 59
Oregon Roadside Council 38, 229
Oregon Silverspot Butterfly 159
Oregon Skyline Trail 48
Oregon State Sanitary Authority 51-53
Oregon State University 34, 35, 50, 98, 100, 131, 157, 175, 191, 201, 204, 206
Oregon Steelheaders 113
Oregon Supreme Court 92, 123
Oregon Water Resources Department 30
Oregon White Oak 164, 171
Oregon Wild 58, 63, 76, 79, 132, 133, 187, 188, 191, 230
Oregon Wilderness Act of 1984 181, 185
Oregon Wilderness Coalition 182
Oregon Wildlife Federation 51
Oregon Wildlife Heritage Foundation 97, 229
Ormsby, S. B. 9
OSPIRG 87, 90, 91, 93, 105
Osprey 151
Oswald West State Park 36, 39
Otter Rock 158
Our Ocean Coalition 158, 230
Owl 155, 194, 197, 198, 200, 206, 230
Owyhee Canyons 214, 228

P

Pacific City 50, 80, 121
Pacific Northwest Planning Commission 62
Pacific Power (and Light) 88, 105
Pacific Railway Survey, 1855 149
Packwood, Robert 57, 82, 142, 143, 147, 176-178
Painted Hills 145
Pamelia Creek 75
Park Division 61
Parks Department 59, 80, 95, 97
Park Service xi, 19, 40, 42, 61, 63, 64, 138, 139, 140, 144, 145, 146, 149, 161, 214
Patterson, Joan 167
Paulina Lake 149, 150
PBDEs 103
Peacock Larkspur 162
Pebble Springs 115
Pelican Bay 31
Pelton Dam 213
People Against Non-Returnables 84
Peregrine Falcon 155
Petterson, Carl 119
PGE. *See* Portland General Electric
Pierce, Walter 211
Pigeon Butte 164
Pilot Rock 155
Pinchot, Gifford 6, 10, 16, 19, 30
Pine marten 151
Pistol River 214
Pittman-Robertson Act 168
Pleistocene 160
Pollution 51, 53, 63, 98
Pollution in Paradise 53
Port Commission, Portland's 120
Portland
 Big Pipes 134-136, 225
 Sewage 52, 134-136
Portland Airport 119, 223
Portland Airport expansion 119
Portland Audubon Society 27, 163, 165, 172, 198, 201, 206, 229
Portland Chamber of Commerce 7
Portland City Club 88
Portland City Council 44, 171

Index

Portland General Electric 105, 107, 113-115, 213
Portland Planning Commission 44
Portland's Heritage Tree Program 171, 224
Portland State University ii, 121
Portland Water Bureau 174, 175
Portland Women's Forum 62
Port Orford 158
Powder River basin 11
Prairie City 7
Primitive Areas 64-66, 69, 74, 77
Prineville 46
Project Foresight 88
Pronghorn 31, 166, 190
Pronghorn Antelope 31
Proposition 6 82
PTA 44
Pulp mills 216, 223
Pumice Grape Fern 151
Puter, Stephen A. D. 46
Puustinen, William 54

Q

Quaintance, Charles 70

R

Rakestraw, Lawrence 9, 10, 34
Range vs. Refuge 31
RARE II 180, 182-184
Reagan, President Ronald 147, 183, 186
Recreation Demonstration Area 40
Recycling 84, 103, 157, 225
Redband Trout 154, 209
Redden, James 203, 205, 225
Redfish Rocks 158
Redmond 42, 154
Red Tree Vole 201
Redwood National Park xi
Reedsport 46, 140
Refuge Recreation Act 168
Remove/Fill Program 59
Renewable Energy Act 104
Renewable Northwest Project 105
Renewable Portfolio Standard 104
Research Natural Areas 159, 160

Reserves
  in Eastern Oregon Mountains 7
  in Western Oregon 14
Richards, W. A. 8, 9
Richmond, Henry 91
Rickreall 164
Rinke, Ken 82
"Rites of Way" 117
River otter 127
Rivers 12, 51, 53, 57, 58, 63, 154, 190, 209, 224
Roaring River Wilderness 190
Roberts, Barbara 212
Roberts, Betty 86
Robert W. Straub State Park 123
Rock Mesa 67, 69, 224
Rocky Mountain Elk 158
Rogue River 2, 57, 58, 69, 77, 132, 134, 223
Rogue River National Forest 2
Roosevelt, President Franklin 33, 211
Roosevelt, President Theodore 7-9, 15, 19, 20, 26-28, 35, 46, 47, 211
Roseburg 21, 34, 55, 222
Rough and Ready Flat 159
Round Butte Dam 213, 227
Ruffed Grouse 155
Russell, Nancy 146-148

S

Safe Drinking Water Act 135
Sahalie Falls 55, 56
Salem 3, 40, 50, 52, 66, 76, 84, 85, 95, 99, 100, 185, 216
Salmon xiii, 59, 132, 134, 141, 184, 188-190, 192, 194, 202-206, 210-213, 216, 225
  Court management of 202
Salmon-Huckleberry (Wilderness) 184, 190
Sands
  Dry 37, 80-83, 121, 123
  Wet 37, 81, 83, 123
San Francisco 27, 70, 120, 135, 140, 161, 248
Santiam River 53, 57, 180, 181, 185, 227
Sargent, Professor Charles S. 6

Sauvie Island 168
Savage Rapids 132
Save the Deschutes 97
Save the McKenzie Association 55, 230
Sawyer, Robert 38, 40, 145
Saylor, John 142
Scenic and Wild Rivers 190
Scenic Areas 93, 146-148, 159, 209, 224, 230
Scenic Waterways 59, 73
Scharff, John 166
School lands 46
Scott, Doug 19, 143, 178, 183, 191
Scott, Harvey 8, 14
Seabirds 25, 26
Sea Lion Caves 139, 140, 141
Sea lions 27
Seaside 37
Seattle xiii, 35, 50, 63, 79, 119, 120, 137, 197, 198, 206
Secretary of Agriculture 33, 47, 73, 130
Seeger, Pete 142
Serena, Al 69
Sewage 52, 89, 134-136, 223
Sewage treatment 223
Sheep 1, 3, 5-7, 12-14, 16, 33
Sheep Rock 146
Sheldon Range 33
Shepherd's Flat Wind Farm 106, 228
Shepperd's Dell Bridge 61
Sherwood 167
Shining Rock Mining Company 186
Sierra Club ii, xi, xii, 5, 19, 42, 57, 67, 70, 76, 89, 91, 93, 105, 108, 129, 140-142, 151-153, 178, 182-184, 193, 194, 196, 197, 215, 217, 229, 248
Sierra Club Legal Defense Fund 197
Siletz Bay 25
Siletz tribe 44
Siltcoos Lake 139, 140
Silver Creek 40
Silver Falls State Park 40, 41, 50, 227
Silverton Hills 100
Simons, David R. 67
Siskiyou 14, 154-156, 177, 188-190, 209, 214, 225-227

Siskiyou Crest 156
Siskiyou National Forest 177, 188, 190
Siskiyou Reserve 14
Sitka Spruce 25
Siuslaw National Forest 15, 128, 129, 139, 159
Siuslaw Reserve 14
Siuslaw River 138, 140
Sixes River, Middle Fork of the 190
Skamania County 146
Skidmore, Owings & Merrill 118
Sky Lakes 185
Sky Lakes Wilderness 181, 227
Slavery, R. G. 2
Smith, Gordon 154, 189
Smith, Margery Hoffman 42
Smith, Robert 149, 151
Smith Rock State Park 42, 227
Snake River 141, 203, 205, 210
Snake River dams 203, 210
Snow Geese 168
Soda Mountain 189, 190
Soda Mountain Wilderness 189
Soda Rock 155
SOM. See Skidmore, Owings & Merrill
Songbirds 167
Southern Pacific Co. 23
South Sister 67
South Slough Estuarine Reserve 223
Spaulding Catchfly 158
Special Areas 67
Special Management Area 148, 150, 151, 209
Spectra Physics 126
Spotted Owl 155, 194, 197, 198, 200, 206, 230
Spotted Owl Reserves 194
Sprague, Charles 76
Spray 12
Spring Basin 188, 190, 230
Stanton, Robert 55
State Land Use Commission 128
State Park System 37, 38
Steel, William G. xi, 4- 6, 16-20, 34, 211
Steens 107, 152, 153, 154, 209, 225, 228, 229
Steens-Alvord Coalition 153

Index

Steens Mountain 152, 154, 209, 228, 229
Steens Mountain Complex 152, 228
Steens Mountain Cooperative Management and Protection Area 154, 225
Steens Paintbrush 152
Steens Thistle 152
St. Helens 113, 149
Stibolt, Tom 167
Straub, Bob 53, 81, 82, 91, 92, 95, 96, 121-123, 130, 141, 180, 211, 215
Straub, Pat 215
Strawberry Mountain 77
Streaked Horned Lark 163
Summer Lake 168, 169, 222, 228
Superfund sites 216
Supreme Court 3, 22, 47, 70, 82, 83, 92, 123, 141, 223
  Oregon 92, 123
Surface mining 110, 210, 224, 229
  Using chemical processing 110

T

Table Rock Wilderness 98
Taft, President William Howard 47, 160
Talbot, Dave 144
The Dalles Dam 213
Thomas Condon Paleontology Center 146
Thomas, Jack Ward 197
Three Arches 222
Three Arch Rocks 25, 28, 226
Three Sisters 64,-68, 77, 79, 176, 181, 185, 227, 230
Three Sisters Wilderness 64, 66-68, 77, 176, 181, 227, 230
  Friends of the 66, 68, 230
Tigard 167
Tillamook 14, 15, 25, 83, 122, 226
Tillamook Reserve 14
Tillamook State Forest 214, 226
Timber and Stone Act 14, 45
Timberline Lodge 42, 48, 50, 209
TMDLs 216
Tongue, Thomas 17
Trails Club of Oregon 43, 229
Trojan nuclear plant 113-116, 225, 227
Trout Mountains 33

Truman, President Harry 25
Tryon Creek State Park 44
Tualatin 167
Tualatin River 162, 167
Tualatin River National Wildlife Refuge 162, 167, 224
Tufted Puffins 26
Tundra Swan 151

U

Udall, Morris 178, 180
Udall, Stewart 122
Ullman, Al 71, 143, 144, 181
Umatilla Basin 103
Umatilla Chemical Depot 217, 227
Umatilla County 12, 13
Umatilla National Forest 12, 13, 177
Umatilla River 204
Umpqua National Forest 2, 14, 15, 177
Umpqua River 55, 57, 222, 227
Underwood, Gilbert Stanley 48
Union County 12
United Nations Biosphere Reserve 159
University of Oregon ii, 35, 66, 96, 139, 145, 178, 248
Upper Klamath Marsh 30, 227
Upper Klamath National Wildlife Refuge 28
Uranium 111, 116
U.S. Public Health Service 52
U.S. Pumice Company 67
U.S. Supreme Court 22, 70. *See also* Supreme Court

V

Vale 110, 111
Valley Oak Savannah 162
Van Duzer Forest Corridor 38
Van Name, Willard 64, 79
Van Strum, Carol 129
Vida 53, 54

W

Waggoner, Don ix, 84
Wagner, David 124

Walden, Greg 134, 153, 154, 189
Waldo, Judge John B. 3-5, 34, 71-73, 211, 227
Waldo Lake 4, 59, 71-73, 227
Wallace, Jr., Henry A. 33
Wallowa City 12
Wallowa County 12
Wallowa Reserve 11
Wallowas 78, 157
Wallowa-Whitman National Forest 11, 12, 70
Wallowa Wilderness 228
Warbler 155
Warm Springs Indian Reservation 74
Warner Reserve 13
Waterfowl xiii, 27, 28, 30, 31, 163, 168, 169, 208, 229
Water Purification and Prevention plan 51
Watts, Lyle 66
Weaver, Jim 73, 176, 180, 181, 184, 185, 212
Webber, Ralph 167
Wenaha Reserve 12
Wenaha-Tucannon Wilderness 177, 181
Wessinger, Bill 85
West Coast Lumbermen's Association 66
Western Forest Industries Association 177
Western Oregon Planning Revision 200
Western Snowy Plover 138
West Eugene Wetlands Plan 127
West, Oswald 9, 15, 36, 37, 45, 60, 83, 123, 211, 226
Wetlands 124-128, 162, 164, 165, 168, 202, 209, 211, 213, 224
  Eugene's 124
Whiteside, Thomas 130
Whitewater Creek 74
Wickiup Plains 67
Wigwam burners 98, 99, 223
Wild and Scenic Rivers 57, 63, 154, 209
Wild and Scenic Rivers system 57
Wilderness Act xi, 67, 74, 76, 176, 180, 181, 185, 191
Wilderness areas ii, xi, 35, 48-50, 58, 64-69, 71, 74, 76-79, 98, 100, 153, 176, 177, 180-183, 185, 188-191, 224, 226-230

Additions to 176
Wilderness Society 153, 183
Wildlife Management Areas 168-170
  Oregon's 168-170
Wildlife refuges xiii, 162, 168, 169, 208, 211, 229
  East of the Cascades 28
Wildlife reserve 154
Wild Rogue 180, 181
Wild Rogue Wilderness 180, 181
Willamette Greenway 94
Willamette Mission State Park 95
Willamette National Forest 2, 176
Willamette River 5153, 93, 94, 96, 118, 134136, 216, 223
Willamette Valley 52, 53, 87, 88, 99, 103, 124, 127, 162, 163, 213, 229
Willamette Valley federal waterfowl refuges 162
Williams, Chuck 146
Williams, Larry ix, 19, 67, 142, 143, 176, 177, 193
Williamson, John 7, 9, 46, 47
Willner, Don 59
Willow Creek 124
Wilson, President Woodrow 211
Wilsonville 89
Wind farms 107
Windmill farm 106
Wind power 106
Winema National Forest 14
Withycombe, Governor 37
WMAs. *See* Wildlife Management Areas
Woahink Lake 139, 140
Wolverton, Charles E. 22
Wolves xiii, 157
Wood, C. E. S. 10
Woolgrowers Association 7
WOPR. *See* Western Oregon Planning Revision
Workman, Jill 153
World Wildlife Fund 156, 214
WPA 48, 50
WPPSS 193
Wren 155
Wu, David 167

Wyden, Ron 101, 102, 154, 184, 189, 212, 225

Y

Yachats 158
Yellow-Bellied Marmots 190
Yellow Rail 30
Yeon, John 48, 62, 146, 148
Yosemite National Park 48

Z

Zigzag Mountain 177
Zilly, Thomas S. 197
Zumwalt Prairie 156, 157, 225, 228, 230

ABOUT THE AUTHOR

Michael McCloskey earned a law degree at the University of Oregon, after growing up in Eugene. Attracted to environmental advocacy, in 1961 he went to work for the national Sierra Club, opening their first field office in the Northwest. After pioneering work for national parks and wilderness, he transferred to San Francisco where he became successively Conservation Director, CEO, and Chairman. He led the Sierra Club during pivotal phases to get framework laws enacted nationally to protect the environment.

After a forty-year career with the Sierra Club, he returned to Oregon upon retirement. Living now in Portland, he chairs the city's Heritage Tree program and writes books, including this one and a memoir (*In the Midst of It*). He also served recently as President of the Federation of Western Outdoor Clubs.